Developing Data-Graph Comprehension in Grades K–8

Third edition

Frances R. Curcio
City University of New York—Queens College
Flushing, New York

Copyright © 2010 by
THE NATIONAL COUNCIL OF TEACHERS OF MATHEMATICS, INC.
1906 Association Drive, Reston, VA 20191-1502
(703) 620-9840; (800) 235-7566; www.nctm.org
All rights reserved

Library of Congress Cataloging-in-Publication Data

Curcio, Frances R.
  Developing data-graph comprehension in grades K-8 / Frances R. Curcio. -- 3rd ed.
    p. cm.
  Includes bibliographical references.
  ISBN 978-0-87353-651-6
  1. Graphic methods--Study and teaching (Elementary) 2. Graphic methods--Study and
teaching (Secondary) I. Title.
  QA90.C86 2010
  372.7--dc22

                                        2010009266

The National Council of Teachers of Mathematics is a public voice of mathematics education, supporting teachers to ensure equitable mathematics learning of the highest quality for all students through vision, leadership, professional development, and research.

Printed in the United States of America

This book is dedicated to the many children and teachers in the New York City Public Schools who recognize the power and importance of the appropriate use of data and graphs. Their contributions to the development of this book are greatly appreciated.

TABLE OF CONTENTS

Preface

Skill in the critical reading of data, which is a component of quantitative literacy, is a necessity in our highly technological society. In particular, processing information presented on the Internet, on television, and in newspapers, magazines, and commercial reports is dependent on a reader's ability to comprehend graphs.

To meet the needs of society, industry, and business, our students must become adept at processing information. With the publication of *Curriculum and Evaluation Standards for School Mathematics,* the National Council of Teachers of Mathematics (NCTM) advocated that children be involved in collecting, organizing, and describing data. Furthermore, it recommended that children construct, read, and interpret graphs as well as analyze trends and predict from the data (NCTM 1989, pp. 54, 109). This position has been reaffirmed in *Principles and Standards for School Mathematics* (NCTM 2000).

This book, a revised and expanded edition of *Developing Data-Graph Comprehension: Elementary and Middle School Activities* (Curcio 1989, 2001), is intended to provide elementary and middle school teachers and teacher educators with practical ideas on incorporating the graph-reading component of quantitative literacy into the instructional program. It can be used to supplement the teachers' editions of grades K–8 textbooks or as an elementary methods text for preservice and in-service teachers. It provides many suggestions for activities that can be used with youngsters in both small-group and large-group instruction. In support of *Principles and Standards for School Mathematics* and *Curriculum Focal Points for Prekindergarten through Grade 8 Mathematics* (NCTM 2006), the activities presented in this book provide teachers with ideas to emphasize exploration, investigation, reasoning, and communication in mathematics. Furthermore, suggestions for integrating technology as a graphing tool are presented in most of the activities.

The material in this book can be used at different grade levels, depending on the learners' prior experiences with collecting and analyzing data. It is important that the data generated or collected by the students be interesting and meaningful to them.

The results of a research study (Curcio 1987) and suggestions of others (Landwehr and Watkins 1986; Nuffield Foundation 1967; Russell 1988) indicate that elementary and middle school students should be actively involved in collecting real-world data to construct their own simple graphs. Learners are encouraged to formulate questions that lead to collecting data (e.g., Who can tie shoes? Which languages do the children in our class speak?), to generate questions about the data collected (e.g., What do the data tell us? What don't the data tell us?), and to verbalize the relationships and patterns observed among the data (e.g., larger than, twice as big as, continuously increasing). In this way, the application of mathematics to the real world may enhance the students' concept development and build and expand their knowledge about mathematics relevant to comprehending the implicit mathematical relationships expressed in graphs.

The development of graph comprehension skills is not meant to be isolated from the rest of the curriculum. Graph reading is not limited to the study of mathematics. Graphs are found in elementary and middle school science and social studies curricula. This book provides ideas for general skills development that can be incorporated across the curriculum. Although in other disciplines graphical representation includes graphs, maps, pictures, and diagrams that may not include numerical information, only graphs that present quantitative information are discussed in this book.

Several modifications have been made in this third edition of *Developing Data-Graph Comprehension in Grades K–8*. Activities related to integrating graphs from the newspaper and data from the Internet have been updated. Suggestions for using and integrating current technology have been included. The application of data and graphs to discussing social justice issues appears in several new activities.

There are eight sections in this book: (1) Graphs—What Are They, and How Are They Used?; (2) Levels of Graph Comprehension; (3) Collecting, Organizing, and Analyzing Data; (4) Inventing and Reinventing Graphical Displays of Data; (5) Constructing Conventional Graphs; (6) Interpreting and Writing about Graphs; (7) Using Technological Tools for Graphing and Analyzing Data; and (8) Classroom Activities. The major portion of this book consists of thirty activities, about one-third of which are new, revised, and updated, designed for immediate classroom use. References to computer software have been updated, and one activity makes reference to using the graphing calculator. Supplementary materials are given in the appendixes. These materials include supplemental graph-reading exercises, topics appropriate for children in grades K–8, instructions for constructing usable aids for teaching graphing skills, different sizes of graph paper, and samples of data collection sheets. The activities and supplemental materials are available for download at www.nctm.org/more4u.

The many teachers and children who participated in the field testing of these activities for the current and previous editions of this book are gratefully acknowledged. In particular, thanks are due to the following individuals, whose positions and school affiliations at the time of the field tests are given: Nicole Francipane, teacher, Queens School of Inquiry, Flushing, New York; Kate Abell, fourth- and fifth-grade teacher, Public School 11, New York City; Susan Folkson, mathematics coordinator, Community School District 25, Bayside, New York; Michelle McCabe, mathematics teacher, Intermediate School 70, New York City; Barbara Nimerofsky, Rossana Perez, and the late J. Lewis McNeese, teachers, Louis Armstrong Middle School, East Elmhurst, New York; and Phyllis Tam, staff developer, New York City Lab School.

Special thanks to Alice F. Artzt and Sydney L. Schwartz, Queens College of the City University of New York, who continue to inspire and challenge my thinking about the development of graph comprehension and for sharing their insights. However, I take full responsibility for any shortcomings in the presentation of this work.

Much of the groundwork for this third edition was established in the previous edition with the assistance of Chris Christopoulos and Peiji Tang. Their contributions are gratefully acknowledged.

Graphs—What Are They, and How Are They Used?

Since the dawn of civilization, pictorial representations and other symbols have been used to record numbers of humans, animals, and inanimate objects on skins, slabs, sticks of wood, and the walls of caves. We can infer the usefulness of this method of recording and communicating information from an ancient Chinese proverb that states, "A picture is worth ten thousand words." Before advanced written forms of communication were developed, people relied on pictorial representations and symbols to record simple statistics. This has proved to be such an efficient and effective method of recording data that it is still used today, although in a quite modified form.

The modern *graph* evolved from the work of the seventeenth-century philosopher and mathematician René Descartes and was further developed by William Playfair, the eighteenth-century English political economist. For the mathematician, a graph is "an invaluable aid in the solution of arithmetic and algebraic problems, the solution of mathematical formulas, and the representation of relationships" (Arkin and Colton 1940, p. 4). For the layperson, a graph helps clarify, organize, and summarize quantitative information found on the Internet and in newspapers, magazines, and advertisements.

Graphs provide a means for communicating and classifying data. Graphs allow for the comparison of data and the display of mathematical relationships that often cannot easily be recognized in numerical form. The traditional, most common forms of graphs found in the media are *picture graphs, bar graphs, line graphs, circle graphs,* and *histograms.*

More recently, some new plotting techniques have been recommended for inclusion in the curriculum by the Joint Committee on the Curriculum in Statistics and Probability of the American Statistical Association and the National Council of Teachers of Mathematics. These techniques include *line plots, stem-and-leaf plots,* and *box plots.*

In the following sections, traditional graph forms and some new plotting techniques are described. Suggestions for presenting each of these can be found in the "Classroom Activities" section that appears later in this book.

Traditional Graph Forms

Picture graphs

The picture graph (also called *pictograph, pictogram, or pictorial graph*) uses representative, uniform pictures to depict quantities of objects or people with respect to labeled axes. It is used when the data are discrete (i.e., not continuous). The ideographs or symbols used must be the same size and shape to avoid misleading the reader (Huff 1954).

In the early grades, children typically draw pictures or use photographs to construct graphs about their birthday months, how they travel to school, and so on. Eventually, the pictures or drawings of real objects are replaced with uniform ideographs. The uniform ideographs may be pictures of real objects (e.g., a stick figure to represent a person or a tooth symbol to represent number of teeth missing), or they may take the form of something more abstract (e.g., a triangle or square).

During the elementary grades, children encounter picture graphs with and without a legend or key. Picture graphs without a legend are easier for children to understand because the ideograph and the item it represents are in a one-to-one correspondence. Once a legend is imposed, the number of objects each ideograph represents must be taken into consideration when interpreting the graph. Fractional parts of ideographs (e.g., one-half of a picture) usually cause some difficulties for children. Ideas for introducing picture graphs to children can be found in Activities 4 and 6–8. More advanced picture graph ideas are in Activities 9 and 15. (See fig. 1a and fig. 1b for examples of picture graphs without and with legends, respectively.)

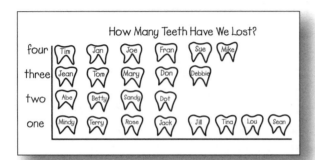

Fig. 1a.

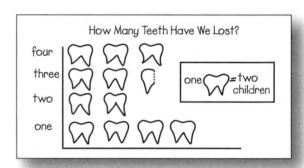

Fig. 1b.

Bar graphs

Set up horizontally or vertically, the bar graph (also called a *bar chart*) allows the reader to compare discrete quantities represented by rectangular bars of uniform width whose heights (or lengths) are proportional to the quantities they represent. The bars are constructed within perpendicular axes that intersect at a common reference point, usually zero. The axes are labeled.

Data presented in picture graphs are usually appropriate for bar graphs. Converting a picture graph to a bar graph is a natural way to help children move from a semiconcrete representation of data to a form that is more abstract.

Ideas for introducing bar graphs can be found in Activities 5, 7, and 8. More advanced bar graph ideas are in Activities 10, 11, 15, 17-20, 25, 26, and 30. See figure 2 for an example of a bar graph of the data in figure 1.

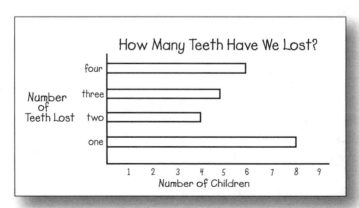

Fig. 2

Double or *multiple bar graphs* are used to compare discrete stratified data (i.e., data collected from particular groups). For example, when asking children to vote for their favorite pets or favorite games to play, the results may be organized according to boys' responses and girls' responses or according to first-graders' responses and second-graders' responses. Double bar graphs are presented in Activities 10, 11, 18–20, 25, and 26. See figure 3 for a sample of this type of graph (based on the data in fig. 1.)

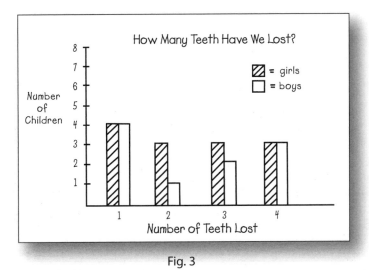

Fig. 3

Line graphs

A line graph (also called a *broken-line graph)* is used to display continuous data. Points are plotted in a plane defined by perpendicular axes to represent change over a period of time or any linear functional relationship. The axes, which are labeled, intersect at a common point, usually zero. The units of division on each axis are equally spaced. The graphed points are connected by straight or broken lines.

Children can keep a record over a period of time of their own height or weight, of the daily average temperature, of the height of a plant, and so on. Ideas for discussing the uses of line graphs are in Activities 21–24 and 27. See figure 4 for an example.

Multiple line graphs are used to compare two sets of continuous data—for example, to compare the heights or weights of two children over a period of time (e.g., four months or one year) or the heights of two (or more) plants over a period of time (e.g., one to two months after planting seeds). Ideas for presenting multiple line graphs are in Activities 23 and 24. See figure 5 for an example.

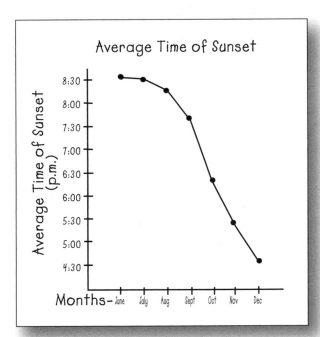

Fig. 4

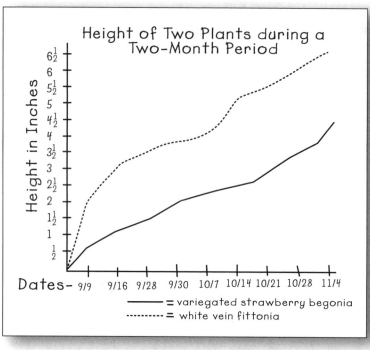

Fig. 5

Circle graphs

The area of the circle graph (also called a *pie graph, pie chart, pie diagram,* or *area graph*) is divided into sectors by radii of the circle. "Each [sector] represents a proportionate part of the whole" (Arkin and Colton 1940, p. 131). The circle graph is used when data indicate a comparison of parts to the whole.

Presenting circle graphs to children is appropriate after they have an understanding of parts of a whole, or fractions. Successful construction of a circle graph is dependent on children's understanding of proportion and their ability to use compasses and protractors. Data for such topics as budget and population characteristics are sometimes displayed in circle graphs. Of all the traditional graph types, the circle graph is the most difficult to construct.

Circle graphs, like bar graphs, are used to display discrete data. Circle graphs are restricted to showing all parts of a whole, and are unlike bar graphs that are viewed as being more useful because they can be used to compare any set of values with the same scale. Furthermore, comparing values close to each other is easier in a bar graph than in a circle graph (David S. Moore, personal communication, 30 October 2009). In response to these limitations, some statisticians and mathematics educators believe that there should be less emphasis on circle graphs in the curriculum (e.g., Susan Friel, personal communication, 3 November 2009; James Landwehr, personal communication, 24 December 1998, 2 November 2009; Albert Shulte, personal communication, 4 January 1999). However, since circle graphs continue to appear in the media, this graph form still has a place in the curriculum. Ideas for discussing circle graphs are in Activities 2, 12–15, and 17. See figure 6 for an example.

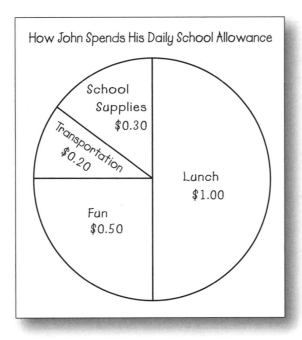

Fig. 6

Histograms

A *histogram* looks somewhat like a bar graph, but unlike a bar graph used to represent discrete data, a histogram is used to represent continuous data organized in intervals, and the bars are contiguous (i.e., adjoining). The width of the base of each bar is uniform. At the base of each bar, the interval, uniform for all bars, is identified. The height of each bar represents the absolute frequency (i.e., the number of data values in the interval) or the relative frequency (i.e., the percent of the data values in the interval; see fig. 7).

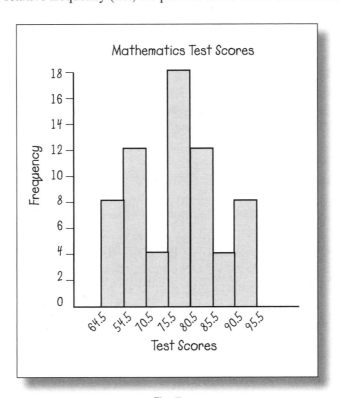

Fig. 7

Constructing a histogram is somewhat like constructing a bar graph in that a set of perpendicular axes is needed. The vertical bars are contiguous, and data intervals are uniform and indicated along the horizontal axis. The axes are labeled. The graph is to have a title. Ideas for discussing histograms are in Activity 28.

Some New Techniques

Following the development of the stem-and-leaf plot and the box plot attributed to John Tukey (1977), a number of authors have suggested that the use of line plots, stem-and-leaf plots, and box plots should be incorporated into the elementary and middle school curriculum (Corwin and Friel 1988; Landwehr and Watkins 1986; Silverman 1988). It has been noted that these new techniques for displaying data have certain advantages over the traditional graphical forms (Landwehr and Watkins 1986).

Line plots

Line plots look like primitive bar graphs, where numerical data are plotted as x's placed above numbers on a number line. A line plot "gives a graphical picture of the relative sizes of numbers, and it helps you to make sure that you aren't missing important information" (Landwehr and Watkins 1986, p. 5). If a bar graph is constructed from grouped data, some of the data may be lost in the grouping; none of the data get lost in a line plot. Usually the number of items plotted does not exceed twenty-five. See figure 8 for an example of a line plot.

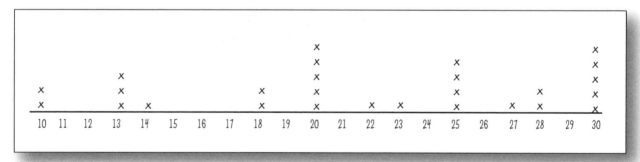

Fig. 8. Estimates for the number of beans in a handful, depicted on a line plot

Stem-and-leaf plots

These plots are characterized by a separation of the digits in numerical data. For example, in a simple stem-and-leaf plot, appropriate for fifth graders, the tens digits are listed in one column and the ones digits are listed in rows next to the respective tens digit (see fig. 9). When rotated ninety degrees counterclockwise, the stem-and-leaf plot resembles a bar graph. This plot "is often better than the bar graph … because it is easier to construct and all the original data values are displayed" (Landwehr and Watkins 1986, p. 9).

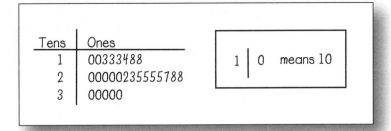

Fig. 9. Estimates for the number of beans in a handful, depicted in a stem-and-leaf plot

Back-to-back stem-and-leaf plots, in which the ones digits of another set of data are attached to the left-hand side of a stem-and-leaf plot, are more difficult for students to read (see fig. 10).

Actual Count		Estimates	
Ones	Tens	Ones	
9999998888	1	00333488	
99553222211111110	2	00000235555788	
	3	00000	

8 | 1 | 0 means
18 actual count
10 estimate

Fig. 10. Actual counts and estimates of beans in a handful, depicted in a back-to-back stem-and-leaf plot

Box plots

Box plots (also referred to as *box-and-whisker plots*) use five summary numbers—the minimum value, the lower quartile, the median, the upper quartile, and the maximum value—and are helpful when analyzing large quantities of data. Although this type of display may be more difficult to construct, it has been used effectively with upper elementary school students.

When using a box plot, "we can no longer spot clusters and gaps, nor can we identify the shape of the distribution as clearly as with line plots or stem-and-leaf plots. However, we are able to focus on the relative positions of different sets of data and thereby compare them more easily" (Landwehr and Watkins 1986, p. 73). See figure 11 for an example of a box plot.

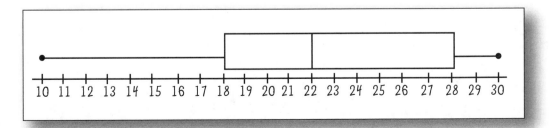

Fig. 11. Estimates for the number of beans in a handful, depicted in a box plot

These three plotting techniques—the line plot, the stem-and-leaf plot, and the box plot—are presented in Activity 20. Teachers are encouraged to use these ideas in other data-gathering and data-analyzing activities.

Levels of Graph Comprehension

Although literal reading of data presented in graphical form is an important component of graph-reading ability, the maximum potential of the graph is realized when the reader is capable of interpreting and generalizing from the data (Kirk, Eggen, and Kauchak 1980). Just as the reading comprehension literature indicates the existence of three levels of comprehension (Pearson and Johnson 1978; Bishop and Bishop 2010), the results of a research study suggest that there are three distinct levels of graph comprehension (Curcio 1987). Regardless of the graph form used, the three levels of graph comprehension are *reading the data, reading between the data,* and *reading beyond the data.*

Reading the Data

This level of comprehension requires a literal reading of the graph. The reader simply "lifts" the facts explicitly stated in the graph, or the information found in the graph title and axes labels, directly from the graph. There is no interpretation at this level. Reading that requires this type of comprehension is a very low-level cognitive task. Questions 1 and 2 for all the supplemental graph-reading activities in Appendix 1 are examples of reading-the-data questions.

Reading between the Data

This level of comprehension includes the interpretation and integration of the data in the graph. It requires the ability to compare quantities (e.g., greater than, tallest, smallest) and the use of other mathematical concepts and skills (e.g., addition, subtraction, multiplication, division) that allow the reader to combine and integrate data and identify the mathematical relationships expressed in the graph.

Reading between the data requires "at least one step of logical or pragmatic inferring necessary to get from the question to the response *and* both question and response are derived from the text" (Pearson and Johnson 1978, p. 161). Questions 3 and 4 for all the graphs in Appendix 1 are examples of reading-between-the-data questions.

Reading beyond the Data

This level of comprehension requires the reader to predict or infer from the data by tapping existing schemata (i.e., background knowledge, knowledge in memory) for information that is neither explicitly nor implicitly stated in the graph. Whereas reading between the data might require that the reader make an inference that is based on the data presented in the graph, reading beyond the data requires that the inference be made on the basis of information in the reader's head, not in the graph. Questions 5 and 6 for all the graphs in Appendix 1 are examples of reading-beyond-the-data questions.

As examples of assessing students' levels of graph comprehension, a section entitled "Questions for Discussion" is included in each of the classroom activities. The questions intended for reading the data (RD), reading between the data

(RBW), and reading beyond the data (RBY) are identified as such. However, the level of comprehension for open-ended questions cannot be determined until a response is given (Pearson and Johnson 1978). It is important to note that children may interpret the questions differently.

In addition to the sample questions provided in each activity, teachers are encouraged to formulate other questions that require reading between and beyond the data. Also, at the end of each question section, a suggestion is made that students think of questions that relate to the graph being analyzed. Examples of alternative assessment tasks can be found in Activities 17 and 24.

The graphs developed by students are not to be treated as "static displays, an end in themselves," but rather the graphs should lead to further questions to enhance the development of critical thinking (Russell 1988, p. 4; Schwartz 2005; Whitin and Whitin 1998). It is hoped that some reading-beyond-the-data questions may lead to further investigation and inquiry (see Activities 1, 3, 4, 8, 17, 20, 23, and 24).

Collecting, Organizing, and Analyzing Data

There are many opportunities to develop ideas about graphs and graphing in elementary and middle school curricula. Such opportunities are not limited to the study of mathematics. A hands-on, minds-on study of science, an inquiry-based approach to social studies, and an examination of language in various stories and texts are a few of the settings that naturally lend themselves to collecting, organizing, and analyzing data.

Collecting Data

It is never too early for children to be involved actively in collecting data to construct their own graphs. Interesting and meaningful activities reflect the experiences of children at their particular grade, age, and ability levels. Children may determine their own topics for generating data, or teachers may offer suggestions (see Appendix 2). Whichever approach is taken, it is important for teachers to determine whether topics are appropriate for the children in their classes.

Throughout the grades, one way to involve children in data collection is to have them identify what they would like to know about their peers or about a topic that might be of interest to them (Schwartz 2005). For example, young children may express an interest in knowing who can tie shoes (Folkson 1996; Whitin 1997), who has a computer (Whitin 1997), or the different languages spoken by the children in the class (Folkson 1996). Older students have expressed an interest in charting, graphing, and comparing weather predictions and the actual weather over a specific time (e.g., two weeks) to determine the percent of time the weather forecast is accurate (Curcio and McNeece 1987). Teachers have found that "the success of child-generated data-collection tasks is based on two criteria: (1) the information to be collected must be useful to the children, and (2) the information must be naturally interesting to them" (Curcio and Folkson 1996, p. 385).

Allowing the children to decide how to collect the data and the form of the response (e.g., responding orally, raising hands, attaching a piece of paper on the chalkboard, casting a ballot, making tallies) contributes to building autonomous learners. The amount of data collected and the data values should be meaningful and manageable for the children.

In the middle grades, the class survey or poll is a way to involve the students in data collection (Newman and Turkel 1985). After discussing with the learners a topic that interests them, allow them to formulate survey questions. Before including the questions in their survey, the students discuss them and come to an agreement about phrasing, relevance, importance, and clarity. As mentioned previously, the amount of data collected and the data values should be meaningful and manageable for the students. Prior to conducting the survey, the students discuss how the information they collect might be organized and analyzed. By conducting a survey, students can appreciate the life of a pollster! Composing and conducting surveys are part of Activities 1, 10–12, 15, 17, 26, and 30.

Another source of data for upper-elementary-grade students is the library. Reference books, newspapers, almanacs, and records and facts books can provide a rich source of data. Prior to referring to such sources, students should formulate questions of interest for which they need to seek answers. In some cases, the data values involved may be very large (i.e., in the billions). It is important that the students have an appreciation and understanding of the large numbers they will encounter.

As more students gain access to the Internet and the World Wide Web, they will be able to use data from various Web sites. For example, national data on births, mortality, and health care are available from the National Center for Health Statistics at www.cdc.gov/nchs. National data concerning education can be found at the Web site of the National Center for Education Statistics (nces.ed.gov). Students can download data and use computer software to construct tables and graphs. As rich and exciting as this new medium may be, some caution is needed. Using the Web as a source of data may be a problem. Care must be taken because "information on the Web is often not evaluated before it is posted and may be biased in one way or another" (Marcovitz 1997, p. 18).

Web access may also contribute to students collaborating on projects and collecting and sharing data with their counterparts around the world (e.g., English and Cudmore 2000; Reese and Monroe 1997; Royer 1997). For example, a forum for designing international projects and displaying students' work can be found at www.kidlink.org (Royer 1997).

Often the results of national surveys are reported in the media. Activities 25, 26, and 28 make reference to such national survey data. If the topic of a national survey is appealing to the students, involve them in conducting a similar survey to collect data locally and then compare the results to the newspaper graph (see Activities 11, 15, and 27). Using the newspaper provides real-world, meaningful data not only for native English-speaking students but also for students who are English-language learners (Olivares 1993). An example of a newspaper article that integrates text and a graphic display can be found in Activity 30.

All the activities in this book have been field-tested and are meant to be interesting and relevant, involving data values that are meaningful and manageable for the students. Teachers may want to make adjustments so that the activities meet the needs of the individual children and classes.

Organizing Data

Once the data are collected, the children become aware of the need to organize the data in a meaningful way so that the data can be interpreted and analyzed. In the very early grades, the number of data items should be countable within determined categories.

In the upper-elementary and middle grades, the children may need to learn tallying techniques. Teachers should not assume that the children know how to tally efficiently. Although tallying is not difficult, the children should be encouraged to discuss how they might go about tallying the data. Some children may have ideas to share with others.

Organizing the data in a chart or table is a very helpful way of getting a picture of the data. The occurrence of certain numbers or patterns will help the students decide how to set up a graph for the data. Getting a feel for the data is necessary in determining which graph form is appropriate for displaying them. It is important to note that tables and charts are often presented in the media without accompanying graphs. However, presenting the data in a graph can often reveal patterns and trends not readily visible in the chart or table. A comparison of data presented in different forms can be found in Activities 15, 17, 20, and 27.

Analyzing Data

Once the data are organized, the students determine whether they are presented adequately for examining patterns, relationships, or trends. Explaining what the data mean explicitly is necessary before examining the implicit relationships between and among the data items. Information presented in a chart or table may need to be converted to a graph to read between and beyond the data. Can any predictions be made? Can any "What if?" questions be asked? Are there any questions that arise from the data collected? Are there any noticeable gaps or holes in the data? Are there any clusters of data? Do there seem to be any outliers? Why might this be so?

In each of the activities in this book, "Questions for Discussion" are meant to guide the students in reading between and beyond the data. If the students are interested, they may want to conduct another survey and construct another graph. They should be encouraged to pursue related projects and report their findings to the class.

Inventing and Reinventing Graphical Displays of Data

Traditionally, graphs have appeared in the elementary and middle school curricula as an end in themselves. That is, the children are *given* a topic, they may be involved in collecting the data, they are *told* the type of conventional graph to construct, and then they are *presented with questions* about the graph that usually involve reading-the-data or reading-between-the-data comprehension. Although this approach may accomplish a teacher's instructional objective, the value of data collection and analysis is not adequately revealed. Instead, a number of things should be made clear to the students.

First, there is a reason to collect data. The reason may be simply that the children have a burning question to ask their peers. Another reason may be to make individual or class decisions. Whatever the reason, it emanates from discussions with the learners and from being sensitive to what interests them (Moersch 1995).

Second, there is a need to communicate the results from the data collection task so that others may use the data for their own purposes and may continue to refer to the visual display for confirmation, clarification, or extrapolation of information. The learners will be challenged to present the data in ways that accurately communicate the message they want to convey.

Third, graphs support data collection and analysis by transforming numerical information in a way that allows for recognizing patterns and trends and facilitating the interpretation and extrapolation of information.

One way to communicate all this to the learners is by allowing them to invent graphical displays using data they have decided to collect. In the early elementary grades, the children may be experimenting with displaying data without having had any prior conventional graphing experiences. In this case, the term *inventing* visual displays is used. In the middle grades, the students who have had prior conventional graphing experiences may be considered to be *reinventing* graphs (diSessa et al. 1991).

Inventing Graphs

Young children who have not had formal instruction in arithmetic have been observed and studied as they created their own problem-solving strategies (Carpenter and Moser 1984) and their own algorithms (Davis 1984; Kamii and Dominick 1997). Based on these observations and the first-grade graphing experiences reported by Whitin, Mills, and O'Keefe (1990), an exploratory study examined the kinds of visual displays created and used by kindergarten children as a means to communicate results of collecting data in order to answer questions of interest to them (Curcio and Folkson 1996; Folkson 1996).

As early as kindergarten, children express interest in collecting data and sharing their results with their peers, and they need to have that interest satisfied (Curcio and Folkson 1996; Folkson 1996). Tasks that involve posing questions and collecting, recording, and analyzing data evolve "from informal discourse and sharing sessions, [as a] result of eavesdropping, listening to children's observations, and prompting them to think about their observations and insights" (Curcio and Folkson 1996, p. 382).

Reading stories aloud sometimes provokes children to pose questions that support data collection. Reading stories helps children appreciate and enjoy the language, the plot, and the illustrations, and sometimes children express interest in examining aspects of the story that the teacher may not have intended. For example, one day while children were enjoying the book *Poor Old Polly* (Melser and Cowley 1980), two children noticed that the words *poor* and *Polly* begin with the same letter as their first names. These observations generated interest in searching for all words in the story beginning with the letter *p* (Curcio and Folkson 1996, p. 384). Working in pairs, the children conducted the search and collected and recorded the data. Some children wrote each word beginning with *p* on paper, circling the words. Other children counted the number of words beginning with *p* and recorded the number on their paper. Jonathan and Michael, two children working together, wrote each word and recorded the number (see fig. 12).

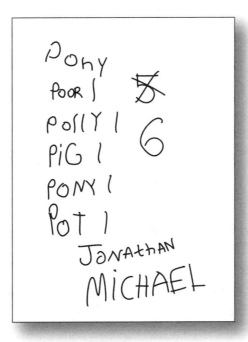

Fig. 12

The boys originally thought there were five words beginning with *p*, until they found *pony* and had to locate space at the top of the list to insert it. They recounted the data and, after crossing out 5, wrote 6 to identify the number of words beginning with *p* in the story. Their work exemplifies reading-the-data comprehension. Jonathan and Michael's visual display in figure 12 does not resemble a rectangular-coordinate grid. But why should it? Traditional, conventional visual displays of data created by adults "do not represent young children's thinking" (Folkson 1996, p. 29).

After reading and enjoying *I Like the Rain* (Belanger 1988a) and *The T-Shirt Song* (Belanger 1988b), the repetitive word *I* became the focus of attention for some children. Rachel wanted to know which book had more *I*'s. Posing this as a question for the children to think about, the teacher elicited the children's estimates. Because the estimates varied, the children wanted to search further. This supports the "estimate, predict, and explore" approach described by Whitin, Mills, and O'Keefe (1990, p. 93). To answer Rachel's question, the children collected and recorded data using such tools as counters, paper, and pencils. Depending on their emerging graphing styles, the children used various recording techniques. The four techniques they used are shown in figure 13.

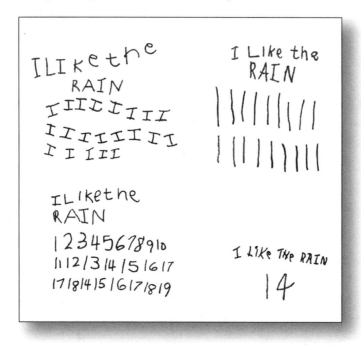

Fig. 13

On the basis of visual displays generated by the kindergarten children (Curcio and Folkson 1996; Folkson 1996) and a sample of the first-graders' work (Whitin, Mills, and O'Keefe 1990), children seem to use various strategies to record their data (Curcio and Folkson 1996, pp. 384–385):

- Some children write the word every time they see it, a tiring and time-consuming effort.

- Others write numbers as they count. This process gives evidence of mastery in counting, numeration, and one-to-one correspondence.

- Some children use tally marks as they count, indicating mastery of one-to-one correspondence and an understanding of abstract representation.

- Some find no need to represent anything on paper except the final number.

Taking a survey may supply useful information for children who need help tying their shoes. When the adults in the room are busy, who can help the children tie their shoes? A dynamic visual display depicting who can tie shoes was created by Olivia, who polled everyone in her kindergarten class and then decided to list only those students who could tie shoes (Folkson 1996). Olivia's visual display in figure 14 was dynamic because as more children were able to tie shoes, Olivia added their names to the visual display, and the display was conveniently located so that children could refer to it when they needed their shoes tied. Referring to Olivia's visual display and reading beyond the data, Poonam said, "There are more children who cannot tie shoes than who can tie shoes."

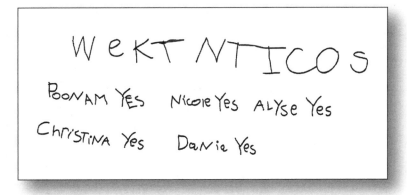

Fig. 14

The children in Queens, New York, the most ethnically diverse community in the nation (Clines 2007), are very conscious of different languages. Many immigrant children enter kindergarten unable to speak English. Some children expressed an interest in determining the languages that the children in the class were able to speak so that if a new child unable to speak English joined the class, the class would know the children who could help interpret for the new child. In comparison to the previous visual displays created by the children, a more sophisticated display was created (see figure 15). To avoid writing the different languages repeatedly, Christina, Jacci, and Olivia used "boxes." One foreign language was assigned to each box, and the names of children who were able to speak the particular languages were written in the appropriate box. To save time and space in writing, the children decided not to include the names of those who could speak only English. This sorting and classification of the data reflect children's recognition of similarities and differences. These skills have been recognized as integral to organizing and graphing data (Baratta-Lorton 1976; Bruni and Silverman 1975).

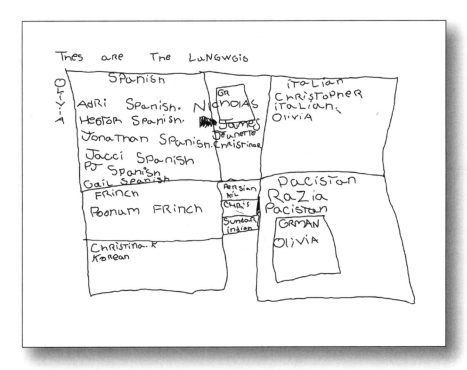

Fig. 15

After the children compiled their data about the languages spoken in class, Olivia's observation exemplified reading between the data: "We found out that a lot of people in class speak Spanish. And Razia is the only one in the class who speaks Pakistani." Another child, Jacci, read the data and read between the data: "I found out that Nicholas speaks Spanish and James speaks Spanish. Persian, I never heard of that—Ali is the only one in the class who speaks Persian."

At the end of the academic year, the kindergarten children expressed an interest in knowing which schools their classmates would attend when they entered first grade. Christina, Nicholas, and Gail collected the data and created a visual display similar to the display about languages (see fig. 16). The names of children planning to attend the same school were listed in the same box. Contrast this display with one created by a child whose "organization skills were still developing" (Folkson 1996, p. 33; see fig. 17).

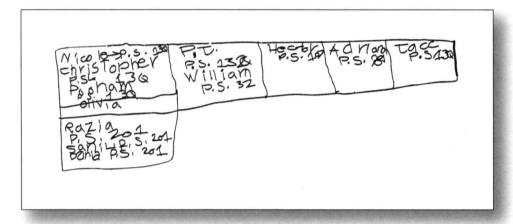

Fig. 16

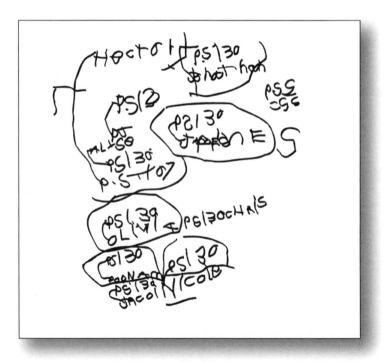

Fig. 17

"Informal discourse among children can happen at any time during the school day—recess, snack time, line-up time—and can become a seed for data-collection and graphing experiences" (Curcio and Folkson 1996, p. 383). One day, after the children had completed lunch and waited for their classmates to use the bathroom, they were sitting opposite their partners with their feet touching (see fig. 18). The children started to compare their shoe sizes. Taking advantage of this teachable moment, the teacher asked them whether they wanted to find out whose foot was the biggest? When the children responded emphatically "yes," the teacher asked them how they could find out. During center time, armed with paper and pencil to record responses, Jeffrey, Rachel, Matthew, and Monique collected shoe-size data. Some children removed their shoes to read the size while others outlined their shoes on paper and compared the lengths of the outlined shapes (Curcio and Folkson 1996). The data collectors decided to hang the outlines on a wall. Reading the data, Jeffrey noticed that Rachel's shoe size was 4. Reading between the data, Olivia noticed that Poonam's shoe was larger than Jonathan's.

Fig. 18

All these examples illustrate how creating visual displays provides the opportunity "for integrating writing, drawing, and mathematics" (Whitin, Mills, and O'Keefe 1990, p. 89). Furthermore, "this approach lays a foundation for children to move from the concrete representation of data to more abstract forms" (Curcio and Folkson 1996, p. 385).

Reinventing Graphs

Upper-elementary and middle school students have been observed and studied reinventing visual displays of data (Curcio and Artzt 1996; diSessa et al. 1991; Tierney and Nemirovsky 1991), some of which is related to graphing functions (diSessa et al. 1991; Tierney and Nemirovsky 1991). Learners involved in inventing, explaining, and discussing their own visual displays are able to communicate levels of conceptualization, visualization, and meaning of phenomena and data that might not be manifested in traditional graph-reading exercises. Middle school students have been challenged to create visual displays of data that they have collected and to describe their thinking about the displays. For example, Brooke, Jason, and Avinesh invented the display in figure 19 to represent the number of letters in their names as they prepared to design a class T-shirt.

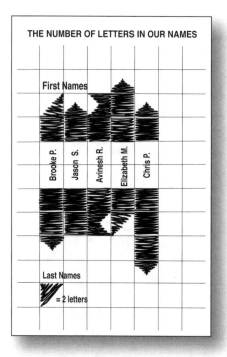

Fig. 19

Brooke, the group recorder, wrote the following:

> This graph shows the number of letters in the first and last names for some of the students in our class. On this graph, each box is worth four letters. We did it this way because when we were graphing the whole class, it did not fit on one big piece of paper, so we tried to shorten it.

"Attempting to solve a problem related to the space on the graph paper, these students demonstrated their knowledge of ratio (i.e., four letters: one box) and fractional parts of a whole (i.e., one-fourth of a box equals one letter) to depict visually the letters in their names. Furthermore, their attempt at producing a display that could be used to compare lengths of first and last names revealed their readiness to discuss elements of a traditional double bar graph" (Curcio and Artzt 1996, p. 671).

From Invention to Convention

Nontraditional, inventive graphing situations open the range of possibilities for eliciting and revealing children's thinking about data and their understandings about the relationships between and beyond the data. As children interact with various media (e.g., Internet, newspapers, television), they encounter conventional graphical forms (e.g., bar graphs, line graphs). How is the transition made from inventions to conventions? Just as comparisons are made between different children's invented visual displays of the same set of data, comparisons can be made between inventions and conventional forms for displaying the same set of data (see Activity 2). By asking the children to describe the similarities and differences between and among the various display formats, the children attend to the characteristics and features of each type of display and identify what information the displays can and cannot supply. The children may be asked to make a list of characteristics, features, and questions that can and cannot be answered on the basis of the displays.

Using graphing software provides the opportunity for students to become familiar with conventional graphical displays. For example, such software as The Graph Club (Stearns 1998), Graphers (Edwards 1996), Data Explorer (Edwards 1998), Microsoft Excel, and such Web sites as illuminations.nctm.org/Activities.aspx?grade = 3&grade = 4&srchstr = graphs and nces.ed.gov/nceskids/createAgraph allow students to enter their own data and explore graphing options. Comparisons of invented graphs and conventional graphs may be made by identifying characteristics and features of the graphs and formulating questions that can and cannot be answered on the basis of the displays (see Activity 2). The students can be asked to explain which graphical form conveys their intended message.

Constructing Conventional Graphs

Graphing ideas develop naturally from activities that involve sorting and classifying (Baratta-Lorton 1976; Bruni and Silverman 1975). As early as kindergarten, children are involved in collecting data about what interests them (e.g., favorite pet, who can tie shoes, the different languages spoken by children in the class, eye color). By using a floor grid (see Appendix 3), the children can create a people graph by lining up according to the appropriate characteristic. For example, all the children with the same hair color or eye color can line up on the floor grid in single file (see Activities 3 and 4).

People graphs lead naturally to block graphs, where children can color, for instance, a pair of eyes drawn on an index card. The index cards can be attached to a block (i.e., same-sized blocks from the block corner), and the children can make piles of similar data blocks (see Activity 4).

People graphs also lead naturally to object graphs, where realia can be used to represent the appropriate data (Choate and Okey 1981). For example, toy cars, toy buses, and doll shoes can be used to represent modes of transportation to school. These toys can be attached to a grid made from oak tag (see Appendix 3), which can then be displayed on a bulletin board or chalkboard (see Activity 6).

On the basis of these primitive graphs (i.e., people graphs, block graphs, and object graphs), picture graphs can be developed. Using uniform ideographs (e.g., pictures of eyes organized by color), the children can identify the one-to-one correspondence represented. By the third grade, the children should be exposed to pictorial representations of simple many-to-one correspondences (for example, one in which two students' votes for favorite pet are represented by one ideograph [see Activity 7]).

Life-sized bar graphs can be created as early as the first grade. The children work in pairs to outline their bodies on poster paper or newsprint, cut out their shapes, and create a height bar graph. To avoid confusing the children with too much data at one time, no more than four height bars are displayed on the same life-sized graph. Following this activity (or instead of it), the children's heights can be represented by lengths of adding-machine tape. Again, no more than four bars are displayed on the same graph (see Activity 5).

Converting a picture graph to a bar graph is a natural way to help the children move from a semiconcrete representation of data to a form that is more abstract. When the teacher recognizes that the children are ready, the picture groupings of one-to-one correspondences on a transparency, chalkboard, or piece of poster paper can be outlined by bars. Then a labeled numerical axis should be inserted and the pictures removed (see Activities 7 and 8).

In the middle grades, the children are involved in discussing the nature of the data collected (i.e., continuous or discrete data), and from that they determine the appropriate type of graph to display the data. If traditional graph forms are used (i.e., picture, bar, line, and circle graphs), the children determine how the data are to be presented and what the title of the graph should be, and then they proceed to plot the data.

As the students compare data displayed in various graph forms, they identify when the data are seemingly distorted or when they seem to be represented fairly. Having access to computers can facilitate graph construction so that the students can focus on comparing and analyzing the graphs rather than on constructing them (see "Using Technological Tools for Graphing and Analyzing Data," p. 25). Also, it is sometimes desirable to display in a circle graph data originally presented in a bar graph. The students examine the similarities and differences of alternative ways of presenting data (see Activity 17).

If nontraditional plots are used (i.e., line, stem-and-leaf, and box plots), students must determine which plot would be most appropriate. If they decide to construct a line plot, they must be able to identify the extremes, proceed to construct a number line, and then place an x for each piece of data in the proper place on the number line. If a stem-and-leaf plot is employed, the children must be able to set up the place-value columns properly and list the data appropriately. If a box plot is to be employed, they should be able to identify the median, the lower and upper quartiles, and the lower and upper extremes (see Activity 20).

Interpreting and Writing about Graphs

Interpreting Graphs

As young children experience the physical creation of a graph, they are involved in interpreting it. Questions that reflect reading the data, reading between the data, and reading beyond the data provide a basis for interpreting and discussing graphs.

Discussion about graphs in the early grades revolves around the language arts (that is, listening, speaking, reading, and writing). While the class listens, individual children are encouraged to talk about the graphs they created, and the teacher (or an adult assistant) records the children's comments. Depending on the children's reading skills, they can read the comments individually or in unison.

These ideas continue through the middle grades. Once graphs are constructed, the children are involved in answering questions "that lead to prediction, interpretations, and to additional questions" (Russell 1988, p. 9). Working in groups of four or five students, they talk about graphs they create, exchange them, and constructively criticize and question their peers.

Writing about Graphs

Activities that provide children with the opportunity to interpret graphs and plots include teacher- and student-formulated questions reflecting different levels of comprehension. On the basis of these questions, children are encouraged to write a paragraph about the graph or plot. Writing about graphs and plots allows the children to clarify their thinking and communicate their interpretation with others. Both the graphs and the written descriptions are shared among the students. The students are encouraged to question and criticize one another constructively. Writing about graphs is a summary feature for each activity in this book.

Using Technological Tools for Graphing and Analyzing Data

With the advent of computer and graphing-calculator technology, teachers are challenged to use these tools to help learners develop skills in accessing, displaying, and analyzing information. Because many children begin school with experience in using the computer and because appropriate graphing software is available for exploration and experimentation (Edwards 1996, 1998; Stearns 1998), the use of the computer in exploratory graphing instruction has found its way into the early grades (Van de Walle 2001, p. 380). Many of the classroom activities in this book include "Using Technology" suggestions for graphing and analyzing data collected by students.

Depending on the students' graphing experiences, the teacher may want to develop ideas using the computer or the graphing calculator instead of having the students construct graphs by hand. However, to understand and appreciate fully the power of the computer to graph data, students must have adequate experience using manipulatives and constructing their own graphs (Edwards 1997; Mathis 1988a; Moersch 1995).

Using the Computer

At the time the teacher believes that students are ready to use the computer, they may or may not be able to recognize the characteristics of a set of data (e.g., discrete versus continuous) that might suggest the use of a particular type of graph. In either event, the students are encouraged to display the data in different graph forms to analyze similarities and differences. Graphs that distort the data and those that represent the data fairly are to be discussed. Whether the intent is to focus on the meaning of the data represented in particular graph forms or to compare advantages and disadvantages of different displays, using the computer in this way will free the students to spend more time interpreting and analyzing graphs instead of constructing them.

To identify graphing software for use with elementary and middle school students, teachers must keep abreast of the development of new products. Software should be consistently reviewed and evaluated to determine whether it is appropriate for use as a tool for graphing data collected by the children.

Stone (1988, p.16) has recommended the consideration of seven factors when reviewing and selecting graphing software tools. A graphing utility should do the following:

1. Provide a choice among a variety of graphing formats and enable alternative representations of a given data set

2. Provide simple, straightforward data entry and editing of information

3. Maximize students' control of labeling, range and number of entries, scaling, and format

4. Make relevant on-screen help readily available

5. Produce clear, easily understandable printouts

6. Allow disk storage of both data set and graph

7. Execute graphs accurately

Although a variety of graphing software is commercially available and new products are likely to become available in the future, currently there is a limited number of good graphing utilities available for use with elementary and middle school students. In particular, table 1 identifies The Graph Club (Stearns 1998), Graphers (Edwards 1996), Data Explorer (Edwards 1998), and Microsoft Excel as having the features that are deemed appropriate for instructional use. For more-detailed reviews of some of these utilities, see Becher (1999), and Curcio (1996, 1998), respectively. The Classroom Activities that refer to using graphing software for which these utilities may be appropriate are also indicated in table 1.

Table 1

Features of Selected Pieces of Software that Support Classroom Activities

Title	Grade Level	Types of Graphs and Plots	Features							Activities
			1[a]	2	3	4	5	6	7	
The Graph Club (Stearns 1998)	K–4	Picture, bar, line, circle, table	+[b]	+	0	+	+	+	+	6–8
Graphers (Edwards 1996)	K–4	Picture, bar, line, circle, table, grid plot, loop (Venn)	+	+	+	+	+	+	+	6–8
Data Explorer (Edwards 1998)	5–9	Venn, bar, multiple bar, line, multiple line, circle, grid, histogram, box plot, stem-and-leaf plot, scatterplot, table	+	+	+	+	+	+	+	7, 8, 10–30
Microsoft Excel (MS Office 2007)	7 and higher	Bar, double bar, line, double line, circle graph, scatterplot, area graph, among others	+	+	+	0	+	+	+	7, 8, 10–30

[a]1. Provide a choice among a variety of graphing formats and enable alternate representations of a given data set
2. Provide simple, straightforward data entry and information editing
3. Maximize students' control of labeling, range, number of entries, scaling, and format
4. Make relevant on-screen help readily available
5. Produce clear, easily understandable printouts
6. Allow disk storage of both data set and graph
7. Execute graphs accurately

[b]+ = Met
0 = Partial

Adapted from Stone (1988) and Mathis (1988b)

The software identified here is offered as a sample of what is available. Database and spreadsheet software packages usually include graphing capabilities, although they may be inappropriate for instructional use in the elementary and middle grades. Teachers can keep abreast of new software developments by examining and reviewing appropriate software on the basis of Stone's (1988) criteria.

Using the Graphing Calculator

The graphing calculator has been recommended for use as early as the middle grades (Lappan et al. 1998). Providing middle school students with graphing calculators is one way to give all learners equal access to technology (Lee 1999).

Graphing calculators with statistics capability offer several important features that support Stone's (1988) criteria. For example, the TI-84 Plus Silver Edition allows students to select from a variety of graph formats typically found in the middle school curriculum, including a scatterplot, line graph, histogram, and box plot. The students may enter and edit data using the "list" feature, and they control the range and scale by adjusting the "window." Using the list feature, data may be stored in the calculator. Using a "graph link," data may be transferred between calculators or from a calculator to a computer, and graphs and data may be printed and stored. When graphs and plots are displayed on the calculator, students must keep in mind the range and scale they establish in the window—no labels appear on the screen. An example of using the graphing calculator may be found in Activity 20. Advances in technology contribute to the frequent appearance of new models of graphing calculators. Teachers are challenged to keep abreast of new developments in this area, also.

Classroom Activities

This section contains thirty field-tested classroom activities. The developmental activities are organized according to grade level. The levels, indicated in the table of contents, are meant to be a guide for planning instruction. Depending on the children's ability, teachers may have to adjust or modify the activities to meet their needs. These activities are also available online at www.nctm.org/more4u.

Although the activities contain many components, they are not intended to be presented to all children, nor are they intended to be accomplished during one lesson. The activities present ideas for developing graphing concepts for children with varying degrees of background knowledge and experiences. The amount of time spent on each component will vary according to the children's ability.

For each activity, there are sections that identify the topic, the graph form to be presented, a graph title, objectives, vocabulary, suggested materials for presenting each graph form, procedures for developing the ideas, suggestions for using the computer as a graphing tool (beginning with Activity 2), questions for discussion, and ideas for summarizing the activity. These sections are repeated in each activity, sometimes verbatim, so that the activities can be used independently.

Many of the topics for the conventional graphing activities in the current edition appeared in the first edition (Curcio 1989). They were determined on the basis of a survey conducted during 1987 of twenty-four elementary and middle school teachers from the New York metropolitan area. Each teacher had more than five years of teaching experience at multiple grade levels. Appropriate grade levels for presenting the graph topics were agreed on by at least 75 percent of the teachers (see Appendix 2 for an extensive list of topics).

Supplemental graph-reading activities are provided in Appendix 1. These are not meant to replace data collection and analysis but rather to supplement the hands-on experiences provided in the classroom activities.

Providing informal graphing experiences may begin with encouraging the children to express an interest in knowing something about their peers, asking them to figure out a way to collect the information from their peers and determine how to record and report what they find out to the class, and allowing them to invent their own visual displays of data (see Activity 1). As children encounter traditional graph forms in newspapers, magazines, and computer software, among other media, they can compare their invented visual displays with traditional graphs and plots. Ideas for making a transition from invented graphs to conventional graphs are suggested in Activity 2. This does not mean that only one lesson is to be dedicated to children's invention of graphic displays and one lesson is needed to make a transition to conventional graphs. Rather, it is recommended that most of the time spent on data collection, recording, and analysis in the early grades (i.e., kindergarten and grade 1) be spent on allowing children to create their own visual displays. It is also recommended that students in grades 2–8 have the opportunity to create their own visual displays. Activity 1 is meant to be an example of how this might be done across the grades.

Most of the graph forms presented in this book are traditional forms (i.e., picture, bar, line, and circle graphs, and histograms). The graph forms presented in

Activities 3–8 and 12 are not formal graphs; they are meant to be informal pre-graphing activities. A more formal treatment of graphing begins in Activity 6, in which using the computer as a graphing tool is presented as an option for teachers. Some activities suggest the presentation of several graph forms so that the children can observe and discuss similarities and differences between and among various graph forms. Ideas for presenting multiple graph forms appear in Activities 4, 6–12, 15, 17, 20, 24, and 27. Teachers decide which graph forms are appropriate as well as when connections between and among multiple forms are to be made.

One activity presents ideas for applying some of the newer plotting techniques. Activity 20 focuses on the line plot, stem-and-leaf plot, back-to-back stem-and-leaf plot, box plot, and multiple box plot. This activity is not meant to be an introduction to these newer techniques but rather to provide a sample application. Teachers are encouraged to apply these to some of the other activities.

Using data and graphs to bring attention to social justice issues is the intent of Activities 25 and 28. Promoting citizens' responsibilities in a democracy requires that teachers provide a platform for students to become aware and to discuss issues and to speak out when they see injustice ("Cutler: It's Up to Us" 2009). Ideas for focusing attention in the study of mathematics on other related social justice issues can be found in Gutstein and Peterson (2006).

In the activity plans, the *Graph title* is usually presented in question form. It is the question that often helps stimulate interest in the activity. After creating a graph, children are expected to identify the graph title. Allow children to select the phrasing of the title, whether the title is displayed on the chalkboard, the overhead projector, document projector, poster paper, or their own graphs.

The *Objectives* present the goals of the different components of the activities with regard to student outcomes. The *Procedure, Questions for discussion*, and *Writing and reading* sections are built around these objectives.

Developing *Vocabulary* is a part of every mathematics lesson. Graphing activities are no exception. Depending on children's ability, the vocabulary words may or may not be familiar to them. New words are used in context and discussed. Familiar words are reviewed. Students are encouraged to use the words as they write about their graphs in the summary phase of the activity.

The *Materials* needed to conduct the activity are listed. However, teachers may feel that they would like to add or delete certain parts of the activity. Also, there may be other materials that have been inadvertently omitted. These should be checked carefully before beginning the activities with the children. Graph paper with different sizes of boxes and a hundredths disk appropriate for different grade levels can be found in Appendixes 5–9. A 24-section circle-graph outline is in Appendix 10.

The *Procedure* presents some opening questions for discussion and an outline of how the activity might proceed. Using children's literature may provide a forum to discuss topics of interest that may lead to graphing tasks. For example, children's books are cited in Activities 2, 3, and 6. Teachers should feel free to deviate from the suggestions and experiment with their own ideas.

Advancements in technology have facilitated displaying quantitative information. Beginning with Activity 2, *Using technology* provides suggestions for graphing and analyzing data collected by students using, for the most part, computers and graphing software. In some (or in most) cases it may be appropriate for students to construct graphs and plots using the computer rather than constructing visual displays by hand. It is recommended, however, that adequate experience using manipulatives and constructing their own invented graphs be provided to help students understand and appreciate fully the power of the computer (Edwards 1997; Mathis 1988a; Moersch 1995). Suggestions for using the graphing calculator are presented in Activity 20.

Questions for discussion lists questions that reflect different levels of comprehension. Abbreviations indicating levels of comprehension are used after the questions: reading the data (RD), reading between the data (RBW), and reading beyond the data (RBY). It is important to note that although questions may be intended for a particular level, the actual level of comprehension cannot really be identified until the student's response is analyzed (Pearson and Johnson 1978). Other graphing activities that contain questions aimed at these levels of comprehension can be found in Appendix 1.

As a means to enhance the children's ability to communicate with mathematics, *Writing and reading* skills are developed and reinforced in the mathematics lesson. Writing about the activity and the graph, reading it aloud or sharing it among the students, and allowing the students to compare the paragraph with the graph afford students the opportunity to clarify their thinking and communicate their interpretations about the graphs with others. This is a part of every graphing activity.

Activity 1

Topic	Students decide
Graph form	Students create
Graph title	Students determine
Objectives	1. To identify topics of interest to be the focus of data-collection tasks
	2. To formulate a question or a series of questions related to the topic(s) of interest to pose to a target audience (e.g., younger students, older students, parents, friends)
	3. To determine ways of collecting and organizing data
	4. To determine ways of summarizing data in a data display to report findings
	5. To interpret data
	6. To answer comprehension questions on the basis of the data display
	7. To write a story based on the data display
Vocabulary	Data, data display, data collection, survey, target audience, organizing data, summarizing data, interpreting data
Materials	Clipboard and paper for each pair of students; pencils, pens, or markers; construction paper; chart paper (or document projector; or overhead projector, transparencies, and markers; or chalkboard and chalk)

Procedure

Ask, "What would you like to know about the children or students in our class?" Elicit ideas from the students, recording the ideas on chart paper (document projector, overhead projector, or chalkboard). When this was posed to kindergarten and grade 1 children at various times during the year (Curcio and Folkson 1996; Folkson 1996; Whitin, Mills, and O'Keefe 1990), the questions posed by the children included—

"Who has the longest foot?"

"What are the different languages the children in the class speak?"

"Do you know how to fix bikes?"

"Does your mom smoke cigarettes?"

Once some ideas are listed, ask the students who would be the target audience to whom the questions would be posed, and in what way knowing the answers to these questions could help us. The students may select a question and work with a partner to determine how to collect the data. They may want to record responses on paper attached to a clipboard. If the students in the class or in other classes in the school are to be polled, provide time for data collection.

Once the data are collected, the students decide how to report their findings to the class. Some students may provide responses in an unorganized, haphazard way, whereas others will make lists, grouping like responses together. Some students may write common responses in boxes. Others may use tally marks or numerals (Curcio and Folkson 1996). Whatever format they use, as the children complete their surveys, they share their work with their peers during a class meeting time. The children are encouraged to ask each other questions and offer suggestions for data presentation and data interpretation. As a means for assessing the students' level of comprehension, teachers may make notes of students' questions, responses to questions, and their interpretations (see pages 17–18).

Questions for discussion

1. "What is your data display about? What title can you give to your data display?" [RD/RBW]

2. "For which [category] are there the most [responses]?" [RBW]

3. "For which [category] are there the fewest [responses]?" [RBW]

4. "Do you think there are more children in all the kindergartens (or first-grade classes) who would be in [a particular category]? Can you answer this question from the data display? Why? Why not? How can we find out the answer to this question?" [RBY] This could lead to collecting information from another kindergarten or first-grade class.

5. "What question can [other students] ask about [student's name] data display?"

Writing and reading

Once the children have responded to the questions, ask them to tell a story about the experience. Write the story as the children dictate it, and then help them read it together.

Activity 2

Topic	Students decide
Graph form	Picture, bar, or circle graph for discrete data, or line graph for continuous data
Graph title	Students determine
Objectives	1. To examine and discuss simple conventional graphs (i.e., picture graphs, bar graphs, circle graphs) from newspapers, magazines, textbooks, or computer software
	2. To describe the features and characteristics of simple conventional graphs from newspapers, magazines, textbooks, or computer software
	3. To compare student-generated data displays with the examples of simple conventional graphs from newspapers, magazines, textbooks, or computer software
	4. To interpret data
	5. To answer comprehension questions on the basis of student-generated data displays and simple conventional graphs
	6. To write a story based on comparing student-generated data displays and simple conventional graphs
Vocabulary	Graph, picture graph (bar graph, circle graph, or line graph), data display
Materials	Simple conventional graphs from newspapers, magazines, or textbooks; computers, graphing software; data collected from students' survey, students' data displays (e.g., from Activity 1); pencils, pens, or markers; chart paper (document projector; or overhead projector, transparencies, and markers; or chalkboard and chalk)

Procedure

A few days prior to this activity, ask students to bring in graphs (i.e., data displays) from newspapers, magazines, or textbooks. (Maintaining an ongoing file of simple conventional graphs would be helpful in the event that the graphs the students find are too complex.)

Ask the students to work in small groups of two or three and examine the graphs they found (or the samples from the file that you distribute) to identify features and characteristics of the graphs. For example, by examining samples of bar graphs, students may identify the L-shaped axes, horizontal or vertical bars, axes' labels, and titles. Referring to a particular graph, ask students to formulate questions that can and cannot be answered. Have students in each group share their work with the whole class, allowing others in the class to question and comment on similarities and differences of their findings.

Using technology

Exploring examples of simple conventional graphs may be accomplished using such software as Graphers (Edwards 1996), Data Explorer (Edwards 1998), and The Graph Club (Stearns 1998).

After spending some time becoming familiar with the software and features of various graphs, ask students to use the data they collected from their surveys to create and print out graphs using the software. Have them compare their data displays (see Activity 1) with the computer-generated graphs of their data, listing similarities and differences.

Questions for discussion

1. "How is your data display the same as the graph you constructed using the computer?" [RBY]

2. "How is your data display different from the graph you constructed using the computer?" [RBY]

3. "What are the characteristics of the graph you created using the computer?" [RBY]

4. "What questions can be answered on the basis of the graph you created using the computer?" [RBY]

5. "What question cannot be answered on the basis of the graph you created using the computer? What further information is needed to answer these questions?" [RBY]

Writing and reading

Once the children have responded to the questions, ask them to write about the graph they constructed using the computer, interpreting the data and comparing the computer-generated graph to the data display created in Activity 1. Some graphing software (e.g., Graphers, Data Explorer, The Graph Club) has a notebook feature that allows students to input their thoughts directly onto the computer. Then the students can print out their graphs and their written descriptions as a record of their work.

Depending on the reading level and interest level of the students, following this activity, the students may enjoy the experiments, puzzles, and games in *Discovering Graph Secrets* (Markle 1997).

Activity 3

Topic	Hair color, hair length, or hairstyle (students decide)
Graph form	People graph
Graph title	What Color (Length or Style of) Hair Do We Have?
Objectives	1. To identify hair color, length, or style
	2. To collect, organize, and interpret data
	3. To create a people graph
	4. To categorize oneself in a people graph according to hair color, hair length, or hairstyle
	5. To answer comprehension questions on the basis of the people graph
	6. To write a story based on the graph
Vocabulary	Graph, people graph; hair colors (review if necessary), hair length or hairstyle (e.g., straight versus curly hair); more, most, fewest
Materials	Floor grid (see Appendix 3); 5–7 pieces of 8 1/2" × 11" paper to use as floor grid labels; paint, crayons, or colored markers to write hair colors, lengths, or styles on floor grid labels; writing tablet

Procedure

Consider reading a children's science-related book about hair with the children (e.g., Goldin 1985). Depending on the children's level of interest, they may want to investigate hair color, hair length (e.g., short, shoulder length, or long), or hairstyle (e.g., straight or curly). Whichever is selected, it is important that not everyone in the class is a member of the same category (i.e., not all children have brown hair).

For example, ask, "What color hair do most of the children in our class have? How can we be sure? What are the different hair colors?" Elicit from the children brown, black, blond, red, and so on. Use appropriate colors of paint or crayons to write the colors on pieces of paper to fit at the base of the floor grid.

Instruct the children with brown hair to line up behind the brown label, those with black hair to line up behind the black label, and so on. Have some of the pupils count the number of children in each row, identifying the number with each hair color.

Questions for discussion

1. "How many children have brown hair? All the children with brown hair raise your right hand." Allow the rest of the children to count those with brown hair and record the number on the chalkboard. Do the same for other colors. [RD]

2. "What title can we give our graph?" Write the title on poster paper, the chalkboard, or on the writing tablet for all the children to see. [RBW]

3. "For which color are there the most children? How do you know this?" [RBW]

4. "For which color are there the fewest children? How do you know this?" [RBW]

5. "Do you think there are more children in all the kindergartens (or first-grade classes) with blond hair or with brown hair? Can you answer this question from the graph? Why? Why not? How can we find out the answer to this question?" [RBY] This could lead to collecting information from another kindergarten or first-grade class.

6. "What question can you ask about our graph?"

Writing and reading

Once the children have responded to the questions, ask them to tell a story about the experience. Write the story as the children dictate it, and then help them read it together.

 # Activity 4

Topic	Eye color (It is important that not everyone in the class is a member of the same category [i.e., not all children have blue eyes].)
Graph form	a. People graph
	b. Block graph (with pictures of eyes colored appropriately)
	c. Picture graph
Graph title	What Color Eyes Do We Have?
Objectives	1. To identify eye color
	2. To collect, organize, and interpret data
	3. To create a people graph, block graph, or picture graph
	4. To categorize oneself in the graphs according to eye color
	5. To answer comprehension questions
	6. To write a story about the graph

Vocabulary	Graph, people graph, block graph, picture graph, eye colors (review if necessary), more, most, fewest
Materials	a. Floor grid (see Appendix 3); ready-made floor grid labels of eye colors (printed in appropriate colors on white paper, 8 1/2" × 11"); construction paper for children to represent their eye colors; writing tablet; marker
	b. Blocks (same size, one for each child); index cards cut to fit the side of the blocks (for children to draw and color their eyes); crayons (to match eye colors); tape (to attach cards to blocks); space on a table or desk to construct a block graph for all the children to see; ready-made block graph labels with eye colors; writing tablet; marker
	c. Construction paper (3" × 5", one piece for each child to represent his or her appropriate eye color); tape (to attach each piece of colored construction paper to a picture graph on the chalkboard); writing tablet; marker

Procedure

Ask, "What color eyes do most of the children in our class have? How can we be sure? What are the different eye colors? (e.g., brown, blue, green, hazel, gray). What color eyes do you have?" For the people graph, have each child select a piece of construction paper to match his or her eye color. Have the children print their names on the paper. For the block graph, distribute index cards and crayons for children to draw and color their eyes. Attach the cards to the blocks.

a. *People graph:* Place the ready-made floor grid labels at the base of the grid (see fig. 4.1). Ask the children with black eyes to line up behind the black-eyes' label, children with blue eyes to line up behind the blue-eyes' label, and so on (see fig. 4.2). Have the children with black eyes raise their pieces of black construction paper. Allow the other children to count those with black eyes and record the number on the chalkboard. Do the same for the other eye colors.

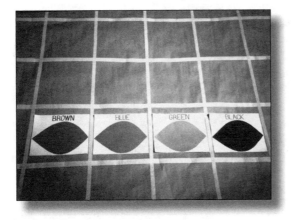

Fig. 4.1

Fig. 4.2

After asking some of the questions listed below, have the children step off the floor grid, leaving their pieces of construction paper in the places where they were standing. Continue with the questioning. Allow the children to sit around the graph, making up a story about the activity (see fig. 4.3). Write the story that the children dictate on the chalkboard or on poster paper (see fig. 4.4).

Fig. 4.3

The Color of Your Eyes

The colors of the eyes of children in our class are: green, black, blue and brown.
Two people have green eyes.
Three people have blue eyes.
Nineteen people have brown eyes
One person has black eyes.

Fig. 4.4

b. *Block graph:* Have the children draw a pair of eyes on an index card and color them to match the color of their eyes. Attach each index card to a block (all blocks should be the same size). Place labels on a table to identify the various eye colors, and have the children stack their blocks in the appropriate pile (see fig. 4.5). After discussing answers to some of the questions below, have the children remove the drawings of their eyes from the blocks (see fig. 4.6); continue with the questioning, at the same time eliciting a story about the activity from the children (see fig. 4.7).

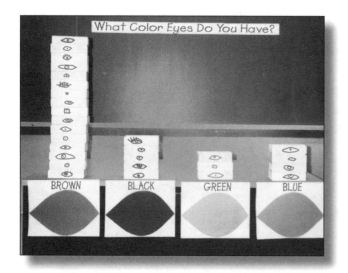

Fig. 4.5

Fig. 4.6

c. *Picture graph:* Allow each child to select a piece of 3" x 5" construction paper to match the color of his or her eyes. Have the children attach their pieces of construction paper in the appropriate row on the chalkboard. Labels of different eye colors are used to identify the colors represented in each row.

> ### What Color Eyes Do You Have?
>
> We drew our eyes and colored them brown, black, green, and blue. We put our drawings on blocks. Then we put the blocks in brown, black, green, and blue sections. We found out that there are the most children with brown eyes and the least children with green eyes. There are a few more kids with black eyes than blue or green eyes. We had 28 blocks showing our eyes. The "eyes" have it!

Fig. 4.7

Questions for discussion

1. "What title can we give our graph?" Write the title on the chalkboard or on paper to post near the graph so all the children can see it. [RBW]

2. "How many children have black eyes? Brown eyes? Blue eyes?" [RD]

3. "What color eyes do most of the children in our class have? How can we be sure?" [RBW]

4. "What color eyes do the fewest children in our class have? How do you know this?" [RBW]

5. "What color hair do most people with brown eyes have?" [RBY]

6. "What color hair do most people with blue eyes have?" [RBY]

7. "Do you think there are more children in our school [or in the kindergartens or first-grade classes] with brown eyes or blue eyes? Why do you think this? Can you answer this question by just looking at the graph? How can we find out?" [RBY] This could lead to collecting information from another kindergarten or first-grade class.

8. "What question can you ask about this graph?"

Writing and reading

Once the questions have been discussed, ask the children to tell a story about the activity. Write their story on the document projector, overhead projector, chalkboard or on poster paper (see fig. 4.7) and encourage the children to read their story aloud.

Activity 5

Topic	Height
Graph form	Life-sized bar graph
	a. Using outlines of children's bodies
	b. Using adding-machine tape

Graph title	How Tall Are We?
Objectives	1. To compare heights by standing next to each other
	2. To construct life-sized models to compare heights
	3. To use adding-machine tape to represent and compare heights
	4. To collect, organize, and interpret data
	5. To answer comprehension questions
	6. To write stories based on the life-sized graphs
Vocabulary	Graph, height, tall, taller, tallest, short, shorter, shortest
Materials	a. Large pieces of drawing paper (or brown wrapping paper) on which to outline the children's bodies; markers; scissors; tape, tacks, or paper clips (to display outlines)
	b. Adding-machine tape to represent the measure of the children's heights; markers; scissors; tape, tacks, or paper clips (to display adding-machine tape)

Procedure

Ask, "Who is the tallest student in our class? How can we be sure?" Students may suggest comparing heights by standing next to one another. Have students in pairs, threes, and fours stand next to one another to compare their heights.

a. Have the children lie on large pieces of drawing paper, newspaper (taped together), or brown wrapping paper while their partners draw outlines of their bodies. Have the children cut out their body outlines and write their names on them. Use three or four outlines at a time to display on a wall or bulletin board as a life-sized graph. Be sure the base of each outline is at the same level for proper comparison. Change the outlines periodically, reviewing the questions below when the changes are made.

b. Have pairs of students use adding-machine tape to represent each child's height, from foot to head. Ask the children to help one another to make sure that the tape does not sag but rather is held taut and perpendicular to the floor. The piece of adding-machine tape can be folded and placed under the child's foot so that the tape can be held taut. Have each child cut the proper length to represent his or her partner's height, and have the children write their names on their tapes. Use three or four lengths of tape at a time to display on a wall or bulletin board as a life-sized graph, being sure that the base of each tape is at the same level for proper comparison. Change the tapes periodically, reviewing the comprehension questions when the changes are made.

Questions for discussion

1. "What would be a good title for this graph?" Write the title above the graph, being sure to include the month and year. [RBW]

2. "What are the names of the children whose heights are represented in the graph?" [RD]

3. "How tall is [insert a name]?" [RD]

4. "Who is the tallest of the students on the graph?" [RBW]

5. "Who is the shortest?" [RBW]

6. "Who do you think is the oldest? Why? Can this be answered directly from the graph?" [RBY]

7. "Who do you think has the smallest shoe size? Why? Can this be answered directly from the graph?" [RBY]

8. "Who do you think weighs the least? Why?" [RBY]

9. "What are some questions you can ask about this graph?"

Writing and reading

After the questions are discussed ask the children to tell a story about the activities. Have them write their own stories, or use the children's words to write the story on the board or poster paper so that all the children can see it. Encourage them to read the story together.

Activity 6

Topic	Travel
Graph form	a. Object graph
	b. Object or person picture graph
	c. Picture graph
Graph title	How Do We Get to School?
Objectives	1. To identify ways of traveling to school
	2. To collect, organize, and interpret data
	3. To use objects to represent the mode of travel and to recognize a one-to-one correspondence between the children and the objects
	4. To recognize that a picture (either of an object or a person) can represent how one travels to school
	5. To recognize that a general object (e.g., a cube) can replace a specific object in a graph
	6. To construct an object graph and a picture graph
	7. To use the computer as a graphing tool
	8. To answer comprehension questions
	9. To write a story based on the graph
Vocabulary	Graph, object graph, picture graph, most
Materials	a. Ready-made chart for object graph (see Appendix 3); model or toy shoes (to represent walkers), buses, cars, boats, and trains; pictures of ways children travel to school (Appendix 4); paper clips (to attach pictures to ready-made chart); tape to attach objects; Unifix or interlocking cubes to represent a generalization of modes of travel; lined paper for writing graph stories; document projector or overhead projector
	b. Ready-made chart for picture graph (see Appendix 3); pictures of children or drawings to represent modes of travel to school (see Appendix 4); paper clips (to attach pictures to ready-made chart); lined paper for writing graph stories; document projector or overhead projector

c. Ready-made chart for picture graph (see Appendix 3); uniform (i.e., same size, shape, and color) geometric shapes cut from construction paper to generalize modes of transportation; paper clips (to attach paper to ready-made chart); lined paper for writing graph stories; document projector or overhead projector; computer and graphing software

Procedure

Ask, "What are the different ways we can travel to school?" List the children's responses on the document projector, overhead projector, chalkboard, or poster paper. Consider reading a children's social studies-related book about how children around the world travel to school (e.g., Baer 1990). Compare the children's list with the various modes of transportation in the story.

a. *Object graph:* Ask individual children the way they travel to school. Have each child pick up a representative model or toy and attach it with tape to the ready-made chart (see fig. 6.1). After discussing the object graph, have the children replace the objects with "interlocking cubes" to represent the modes of transportation.

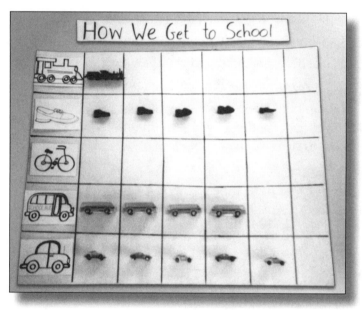

Fig. 6.1

b. *Object or person picture graph:* Ask the children to draw pictures to represent the way they travel to school or to bring in photographs of themselves. Have them attach their drawings or photos to the ready-made chart to indicate which mode of transportation they use to get to school (see fig. 6.2). Appendix 4 is a template for various modes of transportation.

c. *Picture graph:* After discussing the object or person picture graph, have the children replace their pictures with uniform pieces of colored paper (any geometric shape).

Fig. 6.2

Using technology

Students may want to use the data collected for the object graph to construct a picture graph using the computer. For example, figure 6.3 was created using Graphers (Edwards 1996). The class experience story was also entered into the notebook feature of the software.

Fig. 6.3

Questions for discussion

1. "What is a good title for this graph?" [RBW]
2. "How many children walk to school?" [RD]
3. "How many children ride a bike to school?" [RD]
4. "How many children ride the bus to school?" [RD]
5. "How many children are driven to school in a car?" [RD]
6. "How many children take a boat to school?" [RD]
7. "Which is the most popular way for children to travel to school? How do you know? How can you be sure?" [RBW]
8. "How do you think this graph would change if it were raining or snowing?" [RBY]
9. "About how many children live close to school? How do we know this?" [RBY]
10. "What are some questions you can ask about this graph?"

Writing and reading

After the children have discussed the answers to the questions, ask them to tell a story about the activity. Using the children's words, write the story on the document projector, overhead projector, board, or poster paper (see fig. 6.4), and help the children read the story aloud together. As early as possible, encourage the children to write their own stories (see fig. 6.5).

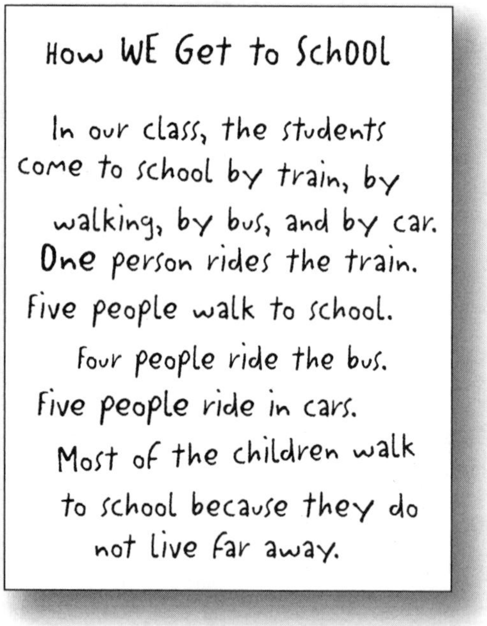

How WE Get to SchOOL

In our class, the students come to school by train, by walking, by bus, and by car. One person rides the train. Five people walk to school.
Four people ride the bus. Five people ride in cars.
Most of the children walk to school because they do not live far away.

Fig. 6.4. Story by class

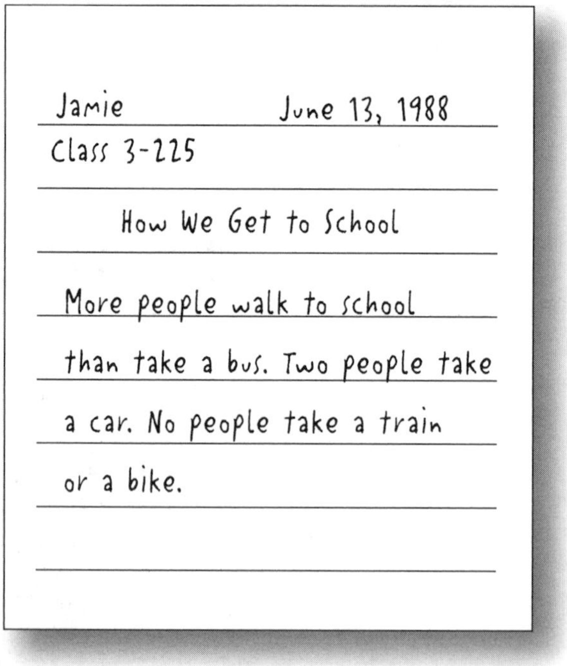

Jamie   June 13, 1988
Class 3-225

How We Get to School

More people walk to school than take a bus. Two people take a car. No people take a train or a bike.

Fig. 6.5. Third grader's experience story about the picture graph

Working together, the class composed a story (see fig. 6.4) that reflects the three levels of comprehension: reading the data (e.g., "One person rides the train."); reading between the data (e.g., "Most of the children walk to school ..."); and reading beyond the data (e.g., "... because they do not live far away."). The third grader's story (see fig. 6.5) reflects two levels of comprehension:

reading the data (e.g., "Two people take a car."); and reading between the data (e.g., "More people walk to school than take a bus."). Analyzing the students' responses to questions and their written interpretations of the graph provides assessment information that may be used to monitor the students' progress in developing graph-comprehension skills.

 Activity 7

Topic	Favorite pet
Graph form	a. Block graph
	b. Picture graph
	c. Bar graph
Graph title	Which Pet Is Our Favorite?
Objectives	1. To identify types of animals suitable for pets
	2. To collect, organize, and interpret data
	3. To categorize one's favorite pet with similar pets
	4. To construct a block graph, picture graph, and bar graph
	5. To relate the results of a picture graph to a bar graph format
	6. To use the computer as a graphing tool
	7. To answer comprehension questions
	8. To write a story about the graph
Vocabulary	Pets, animals, graph, block graph, picture graph, bar graph, best, least, favorite
Materials	a. Blocks (same size, one for each child), index cards cut to fit the side of the blocks, pencils for each child, tape, block graph labels, space on table or desk to construct block graph, writing tablet, marker
	b. Chalkboard, chalk (document projector or overhead projector), white (or light-colored) construction paper, pencils, tape, lined writing paper, computer and graphing software
	c. Chalkboard, chalk, (document projector or overhead projector), lined writing paper, computer and graphing software

Procedure

Ask, "Which kinds of animals do people have as pets?" List the children's responses on the chalkboard. Ask "If you were asked to pick one, which of these would you like best for a pet?"

   a. *Block graph:* Distribute index cards, and ask the children to draw a picture of their favorite pet and attach it to a block. Build a block graph (similar to Activity 4; see fig. 4.5).

   b. *Picture graph:* Distribute construction paper and ask the children to draw a picture of their favorite pet. Ask them to write their names on the back of their drawings and to organize the drawings on the chalkboard according to similarities. Be sure to have labels for the drawings.

   c. *Bar graph:* Using chalk, outline the rows (or columns) of pets posted in the picture graph activity. Insert a horizontal (or vertical) axis indicating the number of each type of pet. Remove the drawings so that the outline of the bar is visible.

Using technology

Children may want to explore creating picture graphs and bar graphs using the computer. For example, figure 7.1 shows a printout from The Graph Club (Stearns 1998). Children click on the type of pet to enter the data. As the picture graph develops, the bar graph develops alongside it. One child noticed, "Each = 1" in small print at the top of the vertical axis of the picture graph, indicating a one-to-one correspondence between the data entered and the ideograph representing the data. When she clicked on this statement, an option to "Choose a new maximum scale" appeared. The student changed the maximum scale from 10 to 20, yielding the screen shown in figure 7.2. The student noticed that the correspondence of ideographs to data items entered changed from a one-to-one correspondence to a one-to-two correspondence (i.e., every ideograph represents two children's votes). Depending on the children's readiness, this provides an opportunity to discuss multiples and ratio.

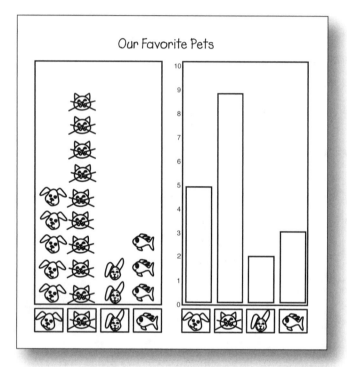

Fig. 7.1

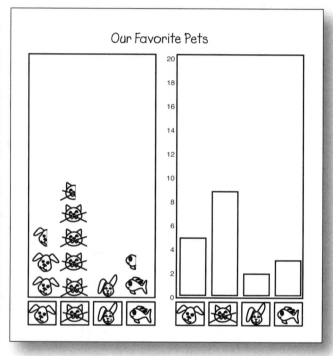

Fig. 7.2

Questions for discussion

1. "What is this graph about? What is a good title for this graph?" Write the title on the graph so that all the children can see it. [RBW]

2. "What are the different animals that we have listed?" [RD]

3. "Which is the class's favorite pet? How do you know?" [RBW]

4. "Which pet is the class's least favorite?" [RBW]

5. "How many of the pets can be trained?" [RBY]

6. "How many of the pets have wings? Which ones are they? How many have four legs? Which ones are they? How many have fins? Which ones are they?" [RBY]

7. "How are the graphs in figure 7.1 and figure 7.2 the same? How are they different?" [RBY]

8. "What does the part of the dog picture in figure 7.2 represent?" [RBW]

9. "Why are there more pet pictures in figure 7.1 than in figure 7.2?" [RBY]

10. "What questions can you ask about this graph?"

Writing and reading

After discussing the answers to the questions, ask the children to tell a story about the graph. Using the children's words, write the story on the document projector, overhead projector, board, or writing tablet so that all the children can see it, and encourage the children to read the story aloud together. As early as possible, encourage children to write their own stories and share them with others.

 # Activity 8

Topic	Favorite ice cream flavors
Graph form	a. Block graph
	b. Picture graph
	c. Bar graph
Graph title	Which Ice Cream Flavor Is Our Favorite?
Objective	1. To identify ice-cream flavors
	2. To collect, organize, and interpret data
	3. To categorize one's favorite ice-cream flavor with similar flavors
	4. To construct a block graph, picture graph, and bar graph
	5. To relate the results of a picture graph to a bar graph
	6. To answer comprehension questions
	7. To write a story about a graph
Vocabulary	Graph, block graph, picture graph, bar graph, ice-cream flavors (review if necessary), favorite, fewest
Materials	a. Blocks (same size, one for each child), index cards cut to fit the side of the blocks, crayons, tape, writing tablet, marker
	b. Chalkboard, white (or light-colored) construction paper, crayons, tape, lined writing paper
	c. Chalkboard, chalk (or document projector or overhead projector), lined writing paper

Procedure

Ask, "What are some of the most popular flavors of ice cream?" List the names dictated by the children. The children may have to decide how to limit the final list of flavors from which to choose the favorite.

a. *Block graph:* Distribute index cards for the children to color to indicate their favorite flavors. Have them attach the index cards to the blocks and stack them according to similarities (similar to Activity 4; see fig. 4.5).

b. *Picture graph:* Distribute construction paper and ask the children to draw and color in their favorite ice cream flavors. Have them organize their drawings on the chalkboard according to similarities. Be sure to have labels for ice cream flavors.

c. *Bar graph:* Using chalk, outline the rows (or columns) of ice cream flavors posted in the picture graph activity. Insert a horizontal (or vertical) axis indicating the number of children who have each flavor as a favorite. Remove the construction paper so that the outline of the bar is visible.

Questions for discussion

1. "What is this graph about? What is a good title for this graph?" Write the title on the graph. [RBW]

2. "Which flavors are listed?" [RD]

3. "How many children like vanilla ice cream the best? Chocolate?" [RBW]

4. "Which flavor is the class's favorite? Why do you think that?" [RBW]

5. "Which flavor is the class's least favorite? Why do you think that?" [RBW]

6. "If we were planning a party, which flavors should we order? How should we decide?" [RBY]

7. "Do you think the other first [or second, and so on] graders like the same flavors that we do? How can we tell? How can we be sure?" [RBY] This may lead into a future activity of polling another class and comparing the results.

8. "What questions can you ask about this graph?"

Writing and reading

After discussing the questions, ask the children to tell a story about the graph. Using the children's words, write the story on the document projector, overhead projector, board, or writing tablet, and encourage them to read it aloud together. As early as possible, encourage children to write their own stories and share them with others.

Activity 9

Topic	Number of children in our families
Graph form	a. Picture graph (using a uniform ideograph and one-to-one correspondence)
	b. Picture graph (using a uniform ideograph and two-to-one correspondence with key)
Graph title	How Many Children Are in Our Families?
Objectives	1. To identify the number of children in one's family
	2. To collect, organize, and interpret data
	3. To categorize the number of children in one's family with other children's responses
	4. To construct a picture graph using symbols so that the number of children and symbols are in one-to-one correspondence and two-to-one correspondence

5. To compare the results of a picture graph in which the number of children and the ideographs are in one-to-one correspondence with one in which the number of children and the ideographs are in two-to-one correspondence

6. To use the computer as a graphing tool

7. To answer comprehension questions on the basis of information in a specific graph

8. To write a story based on a specific graph

Vocabulary   Graph, picture graph, axes, key (or legend)

Materials   Uniform "smiley" faces prepared on Post-It notes (3" x 3", all the same color); one pair of scissors; pencils (one for each child); worksheet for each child (see Appendix 11); chalkboard and chalk (or chart paper and markers, document projector, or overhead transparencies, markers, and projector); computer, graphing software, and printer (optional)

Procedure

Ask the children, "How many brothers and sisters do you have? How many children are in your family?" List their answers on the chalkboard in any unorganized way. Try to identify the smallest number of children and the largest number of children. Prepare a horizontal axis on the chalkboard, labeled "Number of Children." Explain that the number of children in our families will be represented in a picture graph.

Distribute a "smiley" face to each student. Have the students place the Post-It note above the numeral that represents the number of children in their family. As the picture graph is constructed on the chalkboard (chart paper, document projector, or on an overhead transparency), allow the students to construct the graph using Appendix 11. Be sure to label the horizontal axis and identify a title for the graph. Discuss questions 1–5.

Mention that we would like to decrease the number of Post-It notes that we need for this graph: "How can we use fewer notes?" Elicit that we can let each symbol represent more than one student in our class. For example, let each symbol represent two students. "If two symbols represent two students in Graph 1, how many symbols will be needed to represent two students in Graph 2?" Beginning with the first entry of the horizontal axis on the chalkboard, ask the class to convert each column so that one symbol represents two students. For an odd number of students, use a pair of scissors to cut a smiley face in half. Be sure to have the students insert a key (or legend) for each graph. Students may use the right side of Appendix 11 to create the modified picture graph. Be sure to label the horizontal axis and identify a title for the graph. Ask the students to explain the graph (see fig. 9.1). Ask questions 6 and 7.

Fig. 9.1

Using technology

Instead of constructing the picture graphs by hand, or in addition to the above-mentioned activity, students may construct the one-to-one and two-to-one child-to-ideograph picture graphs by using the computer. Select graphing software that has a picture graph capability with the option of creating a legend to adjust the number of items each ideograph represents. Once the children are familiar with how to use the software, allow them to work at computers in groups of two to enter the data, construct the graphs, and compare the computer displays with the graphs they made by hand. Ask the pupils to discuss the similarities and differences between picture graphs that represent one-to-one correspondence and many-to-one correspondence for the same data. If a printer is available, students can print out their graphs. The data and the graphs should be stored on a disk for future reference and use.

Questions for discussion

1. "What is this graph about?" [RD or RBW]
2. "How many students in our class have two children in their families? Three children? Four children?" [RD]
3. "How many children are in the greatest number of families?" [RBW]
4. "How many children are in the fewest number of families?" [RBW]
5. "How many students in our class are the oldest child in their families?" [RBY] Encourage students to compare Graph 1 and Graph 2 to determine whether either one may help answer the question. (Note: Help children realize that neither graph will help answer the question, because neither graph deals with birth order.)
6. "How are the two picture graphs the same? How are they different?" [RBY]
7. "What questions can you ask about these graphs?"

Writing and reading

After the students have discussed the answers to the questions, ask them to write a story about the graph (see fig. 9.2). Ask them to share their work by reading their stories aloud, and encourage them to question and criticize one another constructively.

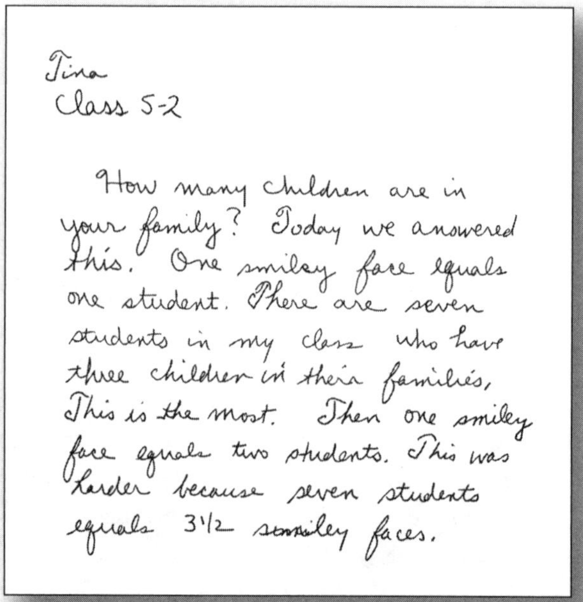

> Tina
> Class 5-2
>
> How many children are in your family? Today we answered this. One smiley face equals one student. There are seven students in my class who have three children in their families. This is the most. Then one smiley face equals two students. This was harder because seven students equals 3½ smiley faces.

Fig. 9.2. A fifth grader's experience story about the graph activity

Activity 10

Topic	Favorite game (board game, card game, athletic or sports game, etc.)
Graph form	a. Bar graph
	b. Double bar graph
Graph title	Which Is Your Favorite Game?
Objectives	1. To identity different types of games
	2. To collect, organize, and interpret data
	3. To categorize one's favorite game with similar responses
	4. To construct a bar graph and a double bar graph
	5. To use the computer as a graphing tool
	6. To answer comprehension questions on the basis of information in a specific graph
	7. To express the meaning of a graph in prose
Vocabulary	Graph, bar graph, double bar graph, legend (or key), horizontal, vertical, survey, poll, data, compare, analyze
Materials	Chalkboard and chalk (document projector or overhead projector, transparencies, and markers); 1-cm graph paper (Appendix 6), pencils, and rulers for each pupil; survey sheet (similar to Appendix 12); computers and graphing software

Procedure

Using a survey form designed by the students, ask them to take a survey of favorite games. The survey can be conducted either in the classroom or as an outside assignment. The pollsters must be sure that they do not collect data from the same student more than once. To construct a double bar graph, they can collect data separately from boys and girls, or from children in different grades. To construct a bar graph, the data can be combined; to construct a double bar graph, the data can be graphed separately with different colors or shadings.

a. *Bar graph:* Once the data have been collected, discuss with the students how to organize a bar graph (if they prefer, a picture graph is also suitable). Distribute 1-cm graph paper and have the students draw a set of axes. Some students may be asked to prepare the graph on the chalkboard (display it on the document projector or on a transparency for the overhead projector). Have the students determine whether the bar graph will be horizontal or vertical. Depending on the number of pieces of data, it might be necessary to use multiples of two or five (or some other number) along the axis labeled "Number of Students." Discuss how to write the numerals along the axis (i.e., stress the meaning of each box and each line). Discuss questions 1–5 below.

b. *Double bar graph:* If the data are collected for boys and girls, or for different grade levels separately, the students can construct a double bar graph. Have the results displayed on the chalkboard, document projector or on an overhead transparency. Distribute 1-cm graph paper, and ask the students to decide whether the bars will be horizontal or vertical. Have them label the axes accordingly. Be sure that the bars have equal widths and that the spacing between "Names of Games" on one of the axes is equal, as well as the spacing between the numerals listed for the "Number of Students." Have the students color or shade the bars and include a legend. They should also identify an appropriate title and insert it on the graph. Discuss questions 6–8 below.

Using technology

Select graphing software that can be used to construct a bar graph and a multiple bar graph. Once the children are familiar with the software, allow them to work at computers in pairs to enter the data, construct the graphs, and compare the computer displays with the graphs made by hand. The software should contain options for adjusting the scale and examining alternative graph forms. Children will need time to explore the options and to discuss the results of using them. If a printer is available, students can print out the graphs to facilitate comparisons. The data and graphs should be saved on a disk for future reference and use.

Questions for discussion

1. "What is this graph about?" [RD or RBW]
2. "What are the different games listed?" [RD]
3. "How many students participated in the survey?" [RBW]
4. "Which game is the most popular?" [RBW]
5. "If you were planning to have a party, how many games would you consider playing at your party? Why? Which games would you select?" [RBY]
6. "Which game is the most popular among the [boys]? [girls]? [fourth graders]? [third graders]?" [RBY]
7. "If you analyze [boys' and girls'] [third graders' and fourth graders'] responses separately, how would you answer question 5?" [RBY]
8. "What questions can you ask about these graphs?"

Writing and reading

After the students have discussed the answers to the questions, ask them to write a description of what the graph means to them. Ask them to share their work by reading their statements aloud, and encourage them to question each other.

Activity 11

Topic	Participation in school musical activities
Graph forms	Bar graph, double bar graph
Graph title	In Which Musical Activities Do We Participate?
Objectives	1. To identify school musical activities
	2. To collect, organize, and interpret data
	3. To categorize school musical activities
	4. To construct a bar graph and a double bar graph
	5. To compare the results of a class (or school) survey with the results of a national survey published in a newspaper

6. To answer comprehension questions on the basis of information in a specific graph

7. To write a letter to the editor of a newspaper related to constructing and comparing graphs

Vocabulary   Graph, bar graph, double bar graph, horizontal, vertical, survey, poll, compare, analyze, percent, representative

Materials   Chalkboard and chalk (document projector, or overhead projector, transparencies, and markers); 1-cm graph paper (Appendix 6) or 1/4" graph paper (Appendix 7), pencils, and rulers for each student; survey sheet (similar to Appendix 12); graph from a national survey (see fig. 11.1)

Procedure

Ask, "Who participates in school musical activities?" "What are the activities that you participate in?" Invite a student to volunteer to record responses on the chalkboard, document projector, or overhead projector. If many sources are suggested, have the students decide how to limit the final list from which they may choose the activity that is specific to them. To compare class results to a national survey, mention the choices to use from the survey (i.e., chorus or choir, band, orchestra, one or more of the three activities; see fig. 11.1).

Fig. 11.1. (Copyright *USA TODAY,* 16 September 2009. Reprinted with permission.)

The survey can be conducted either in the classroom or as an outside assignment. The pollsters must be sure that they do not collect data from the same student more than once. To construct a double bar graph, data collected from boys and girls or from two different age or grade levels will need to be kept separate.

Once the data have been collected, discuss with the students how to organize a double bar graph. Distribute graph paper and have the students draw a set of axes. Do this on the chalkboard, a transparency, or document projector, also. Allow the students to decide whether the double bar graph will be horizontal or vertical. Prepare a set of axes on the chalkboard (on a document projector, or on an overhead transparency), and depending on the orientation of the graph, label the appropriate axis "School Musical Activities." Depending on the number of pieces of data, it might be necessary to use multiples of two or five (or some other number) along the axis labeled "Number of Students." Discuss how to write numerals along the axis, stressing the meaning of each box and each line.

Once the graph is completed, if students are familiar with converting raw data to percentages, have them express the results as percentages so that the graph of the survey data can be compared with the graph in figure 11.1. Otherwise the heights (or lengths) of their bars can be compared with the heights of the musical notes in figure 11.1.

Questions for discussion

1. "What is this graph about?" [RD or RBW]

2. "What are the musical activities represented on the graph?" [RD]

3. "Might there be any other musical activities that are not represented on the graph? If so, what are they?" [RBY]

4. "Why do you think we restricted the number of musical activities in our survey?" [RBY]

5. "Overall, which musical activity do the most students participate in? How do the results in our bar graph compare with those in the graph from the newspaper?" [RBW] If a double bar graph is constructed using data from boys and girls or different age groups, ask, "How can the data in our double bar graph be compared with the data in figure 11.1?" [RBY]

6. "Which musical activity do the most girls participate in [or for a particular age or grade level]? For boys [or for another particular age or grade level]?" [RBW]

7. "What questions should we ask about the results reported in the national survey? Do you think it is representative of the eighth graders in the entire United States? Why? Why not?" [RBY]

8. "Do you think it is fair to compare the class's results with the national survey results? Why? Why not?" [RBY]

Writing and reading

After the students have discussed the answers to the questions, ask them to write a letter to the editor of *USA TODAY*, comparing their graph with the graph in figure 11.1, and pointing out any concerns or questions they have about the published graph. Ask them to share their work by reading their letters aloud. Encourage them to question their peers.

Activity 12

This activity is a result of asking fourth and fifth graders what information they wanted to collect from their peers. Among other topics of interest, the majority of the class decided to construct and conduct a survey about favorite movie stars.

Topic		Favorite celebrities
Graph form	a.	People circle graph (Lovitt and Clarke 1988, pp. 143–49; Van de Walle 2001, p. 372)
	b.	Strip graph (Lovitt and Clarke 1988, pp. 143–49)
	c.	Circle graph
Graph title		Who Is Your Favorite Movie Star?
Objectives	1.	To identify movie stars to be included in a survey
	2.	To collect data by designing a survey
	3.	To organize and interpret data collected in a survey
	4.	To construct a people circle graph
	5.	To construct a strip graph
	6.	To approximate the percent of students selecting each star
	7.	To construct a circle graph using the computer as a graphing tool
	8.	To compare the people circle graph, the strip graph, and the computer-generated circle graph
	9.	To answer comprehension questions on the basis of the information in specific graphs
	10.	To express the meaning of a graph in prose
Vocabulary		People circle graph, strip graph, circle graph, radius, circumference, fractional part, hundredths disk, percent
Materials	a.	Several pieces of string, twine, or cord (each about 2.5 meters long); hundredths disk (see Appendix 9)
	b.	Strips of soft cardboard, oak tag, or adding machine tape for each student or each group of three to four students (e.g., 75 cm x 6.5 cm, for 25 students); markers or crayons; paste, glue, or tape; sheets of butcher paper or brown wrapping paper for each student or each group of three to four students; hundredths disk
	c.	Computers and graphing software

Procedure

a. *People circle graph:* In a whole-group discussion, give the students the opportunity to suggest names of favorite movie stars. Depending on the number of stars identified, the students may need to decide how to limit the number of names to be included on the survey. For example, they may suggest limiting the choices to Leonardo DiCaprio, Eddie Murphy, Will Smith, Jennifer Lopez, Helen Hunt, and Whoopi Goldberg. Students who favor each of the different celebrities meet in groups in different parts of the classroom. Staying together in their groups (i.e., the group of students who select Leonardo DiCaprio), the students stand shoulder-to-shoulder on the circumference of a circle outlined on the classroom floor (see fig. 12.1). All the groups find a place on the circumference of the circle. The ends of long pieces of string are taped to the estimated center of the circle on the floor and the other ends of the string are extended to each point where there is a change in the favorite star. The pieces of string serve as radii of the circle, partitioning the circle into sectors. Discuss questions 1 through 5.

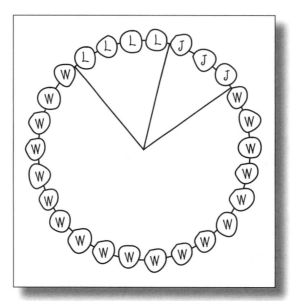

Fig. 12.1. Diagram of students arranged in a circle with letters identifying
movie stars (i.e., L = Leonardo DiCaprio; J = Jennifer Lopez; W = Will Smith)

A hundredths disk is placed on the floor at the center of the circle and the strings mark
off the approximate percents of students selecting each star (Van de Walle 2001, p. 372).
For example, in figure 12.1, approximately 17 percent of the students selected Leonardo
DiCaprio, approximately 10 percent of them selected Jennifer Lopez, and approximately 72
percent of them selected Will Smith. (The actual percents are 15.4, 11.5, and 73.1, respectively.) Although the percents found by using the hundredths disk are approximations,
using the hundredths disk helps students to visualize the meaning of percent. Discuss
questions 6, 7, and 11.

b. *Strip graph:* Using the raw data that are recorded from the people circle graph activity (e.g.,
    four students selected Leonardo DiCaprio, three students selected Jennifer Lopez, and
    nineteen students selected Will Smith), each student is given a strip of adding machine
    tape (e.g., 79 cm x 6.5 cm for each of 26 students). They partition the 79-cm length into
    twenty-six 3-cm spaces, using a pencil to draw lines every 3 cm. The students identify the
    number of contiguous spaces to be assigned to each selected star. For example, four con-
    tiguous spaces are assigned to Leonardo DiCaprio, three contiguous spaces are assigned to
    Jennifer Lopez, and nineteen contiguous spaces are assigned to Will Smith. Some students
    may wish to color-code the spaces; others may wish to use the stars' initials. The extra
    1-cm portion is used as a tab to paste the ends together, making a round band with the
    markings on the outside of the band (see fig. 12.2).

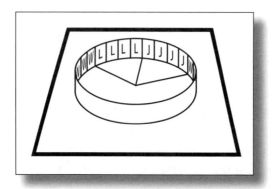

Fig. 12.2. Strip of adding machine tape joined at ends

Once the round bands are made, give the students brown wrapping paper on which to outline the circle, estimate the center, and draw the radii to approximate the sectors representing the data for the three stars (see fig. 12.3). Using their hundredths disk, they approximate the percents of students selecting each star. Discuss questions 1–8 and 11.

c. *Circle graph:* Using the "Explore Category Data" feature of Data Explorer (Edwards 1998), the students enter the class data onto a data sheet (see fig. 12.4). Students may enter the names of their peers in the "Label" column. Individual preferences are listed in the "Variable A" column, identified in this instance as "Stars." Discuss questions 1–5, 7, and 9–11.

Fig. 12.3. Student preparing to outline a circle defined by a strip of adding machine tape with ends joined

Title: Class 4-/5-407's Favorite Stars
Description: Leonardo DiCaprio; Jennifer Lopez; Will Smith

Icon	Label	Variables A
◯		Stars
1	Katherine	Leonardo
2	Rolando	Leonardo
3	Danielle	Leonardo
4	Alex C.	Leonardo
5	Elsa	Jennifer
6	Ana	Jennifer
7	Keyla	Jennifer
8	Ashley	Will
9	Lanaea	Will
10	Iman	Will
11	Jeremy	Will
12	Nicole	Will
13	Alex T.	Will
14	Adam	Will
15	Jenna	Will
16	Gen	Will
17	Maier	Will
18	Chelsea	Will
19	Gregory	Will
20	Adam	Will
21	Temkai	Will
22	Susannah	Will
23	Carter	Will
24	Alison	Will
25	May	Will
26	Lena	Will

Fig. 12.4. Screen created with Data Explorer (Edwards 1998)

Using the computer, the students can print out a table with the frequencies (labeled as "Count"), the fractional part of the whole represented by the portion of the class selecting each star, and the percent equivalent of the fraction (see fig. 12.5).

Some students may wish to construct a circle graph identifying the frequencies (see fig. 12.6), and other students may wish to construct a circle graph identifying the percents (see fig. 12.7). These can be compared to discuss similarities and differences and to formulate questions that can be answered using the different graphs.

Class 4/5-407's Favorite Stars			
Stars	Count	Fraction	Percent
Leonardo	4	4/26	15%
Jennifer	3	3/26	12%
Will	19	19/26	73%
Total	26	26/26	100%

Fig. 12.5. Screen created with Data Explorer (Edwards 1998)

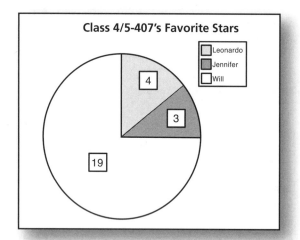

Fig. 12.6. Screen created with Data Explorer (Edwards 1998) highlighting frequencies

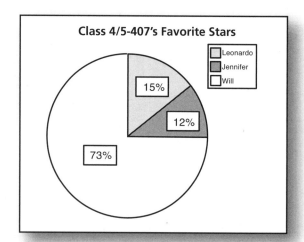

Fig 12.7. Screen created with Data Explorer (Edwards 1998) highlighting percent

Questions for discussion

1. "What is this graph about?" [RD or RBW]

2. "How many students selected Leonardo DiCaprio as their favorite star? Will Smith? Jennifer Lopez?" [RD]

3. "Which star is the class's favorite?" [RBW]

4. "How are Eddie Murphy, Helen Hunt, and Whoopi Goldberg represented on the graph?" [RBY]

5. "What fractional part of the class selected Will Smith? How did you find this out?" [RBY]

6. "How can we read the hundredths disk to determine approximately what percent of the students selected Leonardo DiCaprio? Jennifer Lopez? Will Smith?" [RBY]

7. "If we add the percent of the class selecting each star, what should the sum be? Why?" [RBY]

8. "How are the people circle graph and the adding-machine circle graph the same? How are they different?" [RBY]

9. "How are the people circle graph, the adding machine circle graph, and the circle graph constructed using the computer the same? How are they different?" [RBY]

10. "How did the approximations of the percent of students selecting each star found in the people circle graph compare with what was found using the computer?" [RBY]

11. "What questions can you ask about this graph?"

Writing and reading

After the students have discussed the answers to the questions, ask them to write a description of what the graph means to them. Ask them to share their work by switching papers, data sheets, and graphs. Allow the students to analyze one another's work, encouraging them to question and criticize constructively.

Activity 13

Topic	Daily schedule of activities
Graph form	Circle graph
Graph title	How Do You Spend a Typical Saturday?
Objectives	1. To identify the amount of time spent on different activities during a typical Saturday
	2. To collect, organize, and interpret data
	3. To construct a circle graph on a stenciled circle partitioned into twenty-four congruent sectors
	4. To construct a circle graph using the computer as a graphing tool
	5. To answer comprehension questions on the basis of the information in a specific graph
	6. To express the meaning of a graph in prose
Vocabulary	Graph, circle graph, fractional part, sector
Materials	Data sheet and circle graph outline (Appendix 10) for each student, pencils and crayons, computers and graphing software

Procedure

Ask, "How do you spend a typical Saturday? What are some activities you enjoy doing? How long does each activity take?" To examine activities by the hour, have the students list their activities and the time spent doing them on a data sheet. Once the data have been entered on the sheet, distribute the outline of the circle graph. Students should use each sector as an hour and insert their activities appropriately. (See fig. 13.1.)

Fig. 13.1. Sample of a sixth grader's circle graph

Using technology

Select graphing software that has the capability of constructing circle graphs and expressing results as percents. Once the students are familiar with how to use the software, allow them to work in pairs to enter their data, construct a circle graph, and explore different software options. Have the students print out the graphs they design to facilitate making comparisons. The students may want to store the data and the graphs on a disk for future reference and use.

Questions for discussion

1. "What is this graph about?" [RD or RBW]
2. "How much time do you spend watching television? Eating? Sleeping?" [RD]
3. "On which activity do you spend the least time? The most time?" [RBW]
4. "How might this graph be different for a weekday during the school year? A summer day?" [RBY]
5. "On which activities do you spend a total of more than three hours?" [RD]
6. "On which activities do you spend a total of less than three hours?" [RD]
7. "What fractional part of a Saturday do you spend sleeping? Watching TV? Reading?" [RBY]
8. "After constructing a graph to analyze how you spend a typical Saturday, are there any activities that you think you spend too much time on? Not enough time on? What are they? How would you change these?" [RBY]
9. "What questions can you ask about these graphs?"

Writing and reading

After the students have discussed the answers to the questions, ask them to write a description of what the graph means to them. Ask them to share their work by reading their descriptions aloud or by switching papers among groups. Allow group members to compare the graphs with the written descriptions of them. Allow the students to analyze one another's work, encouraging them to question and criticize constructively (see fig. 13.2).

Fig. 13.2. Sixth graders comparing circle graphs

Activity 14

Topic	Spending daily allowance
Graph form	Circle graph
Graph title	How Do You Spend Your Daily School Allowance?
Objectives	1. To identify the amount of money available for spending on a school day
	2. To collect, organize, and interpret data
	3. To apply knowledge of ratio and proportion for constructing a circle graph
	4. To use a compass and a protractor to construct a circle graph
	5. To use the computer as a graphing tool
	6. To answer comprehension questions on the basis of the information in a specific graph
	7. To express the meaning of a graph in prose
Vocabulary	Graph, circle graph, degrees, compass, protractor, ratio, proportion, percent, fractional part, allowance, typical, central angle
Materials	Overhead projector, transparencies, and markers (or document projector or chalkboard and chalk); review worksheet (see Appendix 14); protractor and compass for the overhead projector (or document projector or for the chalkboard); compasses, protractors, and unlined paper for each student; calculators (to compute fractional parts of 360 degrees); computers and graphing software

Procedure

Ask, "What are some of the items you buy on a typical school day? How much money do you need for school? How much money do your parents or guardians give you for a typical school day? What is an allowance?"

Have the students write down the total amount of money they receive for a daily allowance and how they spend it, listing the items they buy and how much they cost. Have them determine what fractional part of the total allowance is spent for each item (for an example, see fig. 14.1).

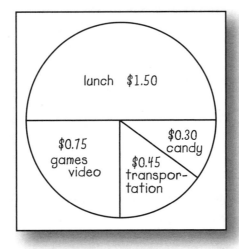

Fig.14.1. Sample of a seventh
grader's circle graph

(If students need to review fractional parts of a circle and determine the number of degrees in certain fractional parts of a circle, consider spending time on an activity similar to Appendix 14 prior to constructing a circle graph. Students may also need some help in using and reading a protractor.)

Distribute unlined paper, compasses, and protractors to the students. (If necessary, review safety procedures for using a compass.) Review the characteristics of a circle (e.g., it contains 360 degrees; it is a set of points equidistant from one point, called the center). Have the students set a radius on their compasses (e.g., 3 inches or approximately 7.5 cm) and draw a circle. Ask such questions as, "If we want to draw one-fourth of the circle, how many degrees would we mark off? [90 degrees] If we want to draw one-half of the circle, how many degrees would we mark off? [180 degrees] If we want to draw one-third of the circle, how many degrees would we mark off? [120 degrees] How can we calculate the number of degrees to mark off when we know the fractional part of the circle that we want to draw?" [Multiply 360 degrees by the fraction or make a proportion; see below.]

"Look at the first item on your list of things that you buy using your daily school allowance. What fractional part of your allowance do you use to buy it?" Have the students multiply 360 degrees by this fraction to compute the number of degrees to mark off in the circle. Review how to construct central angles in the circle with a protractor. Have the students complete their circle graphs by computing fractional parts of the circle and constructing the proper central angles. Have them record the numbers of degrees for the fractional parts on the side of the graph. Students should indicate the item and the amount of money for each item (or the fractional part or percent of the total allowance) in each sector of the circle graph. Be sure to have the students give a title for the graph.

Some students may be able to compute the number of degrees in the central angles to be constructed by setting up a proportion. For example, for lunch (in fig. 14.1),

$$\frac{1.50}{3.00} = \frac{x}{360}.$$

Using technology

Select graphing software that provides options for constructing a circle graph and expressing the results as equivalent fractional parts or percents. Once the students are familiar with how to use the software, allow them to work in pairs to enter their data, construct a circle graph, and explore the different software options. Ask the students to represent the data in different graph forms and describe whether the different forms are appropriate, explaining the criteria they used. To facilitate making comparisons, the students may want to print out their graphs. The data and the graphs could be saved on a disk for future reference and use.

Questions for discussion

1. "What is this graph about?" [RD or RBW]
2. "How much money do you spend for lunch? For transportation?" [RD]
3. "For which item do you spend [insert amount of money]?" [RD]
4. "What part of your allowance do you spend on lunch? Transportation?" [RBW]
5. "How can you budget your money so that you can save some [or more] of it? What nonessential items can you eliminate from your shopping list to save money? Would you want to eliminate them?" [RBY]

6. "If you were given an extra dollar a day, what would you do with it? How would this change the graph?" [RBY]

7. "During a five-day school week, about what part of your five-day allowance is spent on lunch? How does this compare with the part of your daily allowance that you spend on lunch?" [RBY]

8. "Think of a question you can ask about this graph."

Writing and reading

After the students have discussed the answers to the questions, ask them to write a description of what the graph means to them, and have them share their work by switching papers and graphs. Allow them to analyze one another's work, encouraging them to question and criticize constructively.

Activity 15

Topic	Favorite cookies
Graph forms	a. Picture graph (1-to-1 correspondence)
	b. Picture graph (2-to-1 correspondence)
	c. Bar graph
	d. Circle graph
Graph title	What Is Your Favorite Cookie?
Objectives	1. To identify types of cookies
	2. To collect, organize, and interpret data
	3. To categorize data related to types of cookies
	4. To construct a picture graph, bar graph, and/or circle graph
	5. To compare the results of a picture graph in which the number of students and the ideographs are in one-to-one correspondence, with one having two-to-one correspondence
	6. To use the computer as a graphing tool
	7. To compare the results of the class survey, displayed in a circle graph and employing the use of percents, with the results of a national survey published in a newspaper
	8. To answer comprehension questions on the basis of information in a specific graph
	9. To write a letter to the editor of a newspaper related to constructing and comparing graphs
Vocabulary	Graph, picture graph, 1-to-1 correspondence, 2-to-1 correspondence, bar graph, circle graph, axes, key (or legend), percent, representative
Materials	a. Uniform pictures of circular cookie shapes, made from a stencil or on a photocopy machine; chalkboard, document projector, or overhead projector to display symbols (i.e., pictures of cookies); tape; 1-cm graph paper (Appendix 6) for each child; pencils and rulers for each child; computers and graphing software

b. Same as (a), scissors

c. 1/4" graph paper (Appendix 7), pencils, and rulers for each child; computers and graphing software

d. Unlined paper, compasses, rulers, calculators, computers and graphing software

Procedure

Ask, "What is your favorite cookie?" Invite a volunteer to record students' responses on the chalkboard, document projector, or overhead projector. To compare class results to a national survey, suggest that students select the favorites similar to those in a national survey reported in the newspaper (i.e., chocolate chip, peanut butter, oatmeal, sugar/shortbread, and other; see fig. 15.1). Prepare a set of axes on the chalkboard, document projector, or overhead projector and label the vertical axis "Favorite Cookies" and horizontal axis "Number of Students."

Fig. 15.1. Graph of "What is your favorite cookie?"
(Copyright *USA TODAY,* 14 October 2009. Reprinted
with permission.)

a. *Picture graph (one-to-one correspondence):* Give a picture of a cookie to each student. Have each pupil tape the picture in the row that identifies the cookie that he or she likes the best. Distribute 1-cm graph paper so that the students can construct the graph, drawing in the cookies (e.g., circular symbols). Discuss a possible title and insert it on the graph. Discuss questions 1–6 below.

b. *Picture graph (two-to-one correspondence):* Similar to Activity 9, mention that we would like to decrease the number of symbols used in this graph: "How can fewer symbols be used?" Elicit that we can let each symbol represent more than one student in the class. For example, let each symbol represent two students. "If one symbol represents one student in our first graph, how many symbols will be needed to represent two students in this graph?" [Remember, one symbol represents two pupils.] "How many symbols will be needed to represent four pupils?" "How many symbols will be needed to represent five pupils?"

Beginning with the top entry of the vertical axis on the chalkboard, ask the class to convert each row so that one symbol represents two students. For an odd number of students, use a pair of scissors to cut a picture of a cookie in half. Distribute more 1-cm graph paper, and have the students convert the one-to-one picture graph to a two-to-one picture graph. Be sure that they insert a key (or legend) for each graph. Ask them to explain each graph. Discuss question 7 below.

c. *Bar graph:* Distribute 1/4" graph paper, rulers, and pencils. On the chalkboard or document projector, list the following types of cookies: chocolate chip, peanut butter, oatmeal, sugar/shortbread, and other (from the national survey reported in fig. 15.1). Ask the students to select the type of cookie they like the best. Use tally marks on the chalkboard, document projector, or overhead projector to indicate the number of students who select each cookie. Ask the students to construct a bar graph from these data. Be sure that they label each axis and insert an appropriate title for the graph. Depending on students' facility with converting raw data to percentages, have students express the data in percentages.

d. *Circle graph:* Depending on whether the students decide to have the survey open-ended (i.e., allow respondents to state their own favorite cookie type) or closed (i.e., allow a limited number of choices from which respondents can select), ask the students to collect and analyze the data. If students decide to conduct an open-ended survey, it may not be possible to compare the results with the national survey reported in figure 15.1. A poll can be conducted by students and tally marks can be displayed on the chalkboard, document projector, or overhead projector. A people circle graph or strip graph may be constructed (see Activity 12), and students can estimate the percentages of students who selected different types of cookies as their favorite.

If students are ready to construct a circle graph by hand, distribute blank, unlined paper, protractors, compasses, rulers, and pencils, and have the students construct a circle graph, using proportions to find the central angle of the circle to represent the portion of the circle that represents each favorite type of cookie. For example,

$$\frac{\text{No. of students selecting chocolate chip}}{\text{Total no. of students}} = \frac{\text{No. of degrees in the sector of the circle representing chocolate chip cookies}}{360 \text{ degrees}}$$

$$\frac{15}{30} = \frac{x}{360}$$

$$x = 180 \text{ degrees.}$$

Also, make sure they insert a title on the graph. Have students compare their results with the results of the national survey (see fig. 15.1). Discuss question 8 below.

Using technology

The graphing software selected should be capable of displaying the graph forms developed in this activity. Once the children are familiar with the software, allow them to work at the computers in groups of two to enter the data, construct the graphs, and compare the computer displays with the graphs they made by hand. The software should contain options for adjusting the scale and examining alternative graph forms (see Activity 7). The students will need time to explore these options and to discuss the results of using them. Ask the students to examine the similarities and differences among the picture graph, bar graph, and circle graph and to experiment with different scales.

To facilitate a comparison of the class's data with those from the national survey, use the computer to convert the bar graph to a circle graph if the software provides percent equivalents.

Ask students to print out graphs so that they can be compared and discussed. The data and the graphs should be stored on a disk for future reference and use.

Questions for discussion

1.   "What is this graph about?" [RD or RBW]
2.   "How many students like chocolate chip cookies the best? Peanut butter? Oatmeal?" [RD]
3.   "Which is the class's favorite cookie? How do you know this?" [RBW]
4.   "Which is the class's least favorite of the four choices of cookies?" [RBW]
5.   "Who were the respondents supplying the data in figure 15.1?" [RD] "How do the results of the data collected from students in our class compare with the results in figure 15.1?" [RBW] "If you surveyed adults about their favorite cookie rather than students, how do you expect the results would compare with the graph in figure 15.1 and the graph with student data?" [RBY]
6.   "What questions can you ask about this (these) graph(s)?"

For activities (a) and (b), ask—

7.   "How are the two picture graphs the same? How are they different?" [RBY]

For activity (c), ask—

8.   "How do our class results compare with the results of the national survey? What questions should we ask about the results reported in the national survey? Do you think the graph is representative of the entire U.S. population? Why? Why not? Do you think it is fair to compare the two graphs? Why? Why not?" [RBY]
9.   "In figure 15.1, how does the missing part of the cookie being eaten affect the representation of 53 percent of the respondents who selected chocolate chip as their favorite cookie? How might the newspaper artist fix this problem?" [RBY]

Writing and reading

After the students have discussed the answers to the questions, ask them to write a letter to the editor of *USA TODAY*, comparing their graph with the graph in figure 15.1, and pointing out any concerns or questions they have about the published graph. Ask them to share their work by reading their letters aloud. Encourage them to question their peers.

Activity 16

Topic	Height
Graph form	Bar graph
Graph title	How Tall Are You?
Objectives	1. To determine one's height by measuring in metric units
	2. To collect, organize, and interpret data
	3. To compare the heights of students in small groups (i.e., 4–5 students in a group)
	4. To construct a bar graph
	5. To use the computer as a graphing tool
	6. To answer comprehension questions on the basis of information in a specific graph
	7. To express the meaning of a graph in prose
Vocabulary	Graph, bar graph, height, centimeters, meters, metric tape measure, compare
Materials	Approximately ten 1-meter tape measures (two for each group of four or five students) attached to wall; recording sheets (similar to Appendix 13); 1-cm graph paper (Appendix 6) or 1/4" graph paper (Appendix 7), pencils, and rulers for each student; computers and graphing software

Procedure

Ask, "Who is the tallest in our class? How can we be sure?" [By standing next to one another; by measuring heights.] Allow students to work in groups of four or five to measure one another's height and construct a four- or five-item bar graph. Attach two tape measures to the wall in several places around the classroom, end-to-end, to measure students' heights. Review how to read a tape measure and how to add on the amount from the second tape measure. Show the students doing the measuring how to use rulers as a guide, placing them gently on the head of the student being measured (see fig. 16.1). Each student should record the heights of all the members in his or her group.

Fig. 16.1

Once the data have been collected, ask students in each group to discuss how to organize the bar graph. Lead them to suggest that they should organize the bars vertically so that they visually represent height. Distribute graph paper and have the students draw and label a set of axes. Allow students to decide what scale to use so that everyone's height can fit on the graph. Remind the students that equal spacing between bars and between numerals listed along the vertical axis is necessary. Have students choose a title and insert it on the graph.

Using technology

Once their heights have been recorded, have the students work in pairs at the computer to enter their group's data, construct a bar graph, and explore how the graph changes as different scales and alternative graph forms are used. Have the students report the results of their explorations to the rest of the class. Discuss graphs that seemingly distort the data and those that represent the data fairly. Have the students print out their graphs and save the data and the graphs on a disk for future reference and use.

Questions for discussion

1. "What is this graph about?" [RD or RBW]
2. "How tall is [insert name]?" [RD]
3. "Who is the tallest in your group?" [RBW]
4. "Who is the shortest in your group?" [For some children, this may be a sensitive question.] [RBW]
5. "How much taller is [insert name] than [insert name]?" [RBW]
6. "From the information in this graph, can you determine who weighs the most?" [This, too, may be a sensitive question.] "Why did you answer that way?" [RBY]
7. "From the information in the graph, who do you think wears the largest shoe? The smallest?" [RBY]
8. "Think of a question you can ask about this graph."

Writing and reading

After the students have discussed the answers to the questions, ask them to write a description of what the graph means to them. Ask them to share their work by reading aloud what they have written or by switching papers among groups. Allow the group members to compare the graphs with the written description of them. Allow the students to ask one another questions about their written description.

 # Activity 17

Topic	Birthday months (Because some students may not "celebrate" birthdays, the types of questions in this activity, and perhaps the topic, may be inappropriate. Use discretion on the basis of the cultural backgrounds of the students in the class.)
Graph form	Table, bar graph, and circle graph
Graph title	The Number of Children in Our Class Born during Each Month of the Year

Objectives	1. To determine by taking a survey the months of the year in which the students were born
	2. To collect, organize, and interpret data, manually or with a computer (if available)
	3. To organize data in a table
	4. To use the computer (if available) to translate data from a table to a bar graph and a circle graph
	5. To answer comprehension questions on the basis of the displays
	6. To analyze similarities and differences among the displays
	7. To formulate questions that can be answered using the different displays
Vocabulary	Survey, table, bar graph, circle graph
Materials	Computer and graphing software (if available); Appendix 15 (optional); unlined paper, graph paper, protractors, compasses, pencils, and rulers for each student (if the displays are constructed by hand)

Procedure

Ask, "To plan for our monthly recognition of birthdays, during which month of the year do you think the most students in our class were born? How can we find out?" Elicit ideas from the students (e.g., take a survey; see fig. 17.1). "What are the different ways we can display the data we collect?" [e.g., table, picture graph, bar graph, circle graph]. Allow students to conduct a survey and display the data. Depending on the availability of computers and appropriate graphing software, students may work directly on the computer, or they may construct the displays by hand. Appendix 15 may be used as additional assessment tasks related to this activity.

During which month were you born?

We want to know the months of the year the students in class 4/5-307 were born.

1. During which month were you born?

- ◯ January
- ◯ February
- ◯ March
- ◯ April
- ◯ May
- ◯ June
- ◯ July
- ◯ August
- ◯ September
- ◯ October
- ◯ November
- ◯ December

Fig. 17.1. Students' survey queston using Data Explorer
(Edwards 1998)

Using technology

Select graphing software that allows students to enter survey data. Depending on the software, the students may want to enter individual data directly or collect and then enter the data. For example, Data Explorer has a "Survey" feature that allows the students to enter individual data directly. See figures 17.1–17.5 for samples of the computer printouts.

Title: During Which Month Were You Born?
Description: We want to know the months of the year in which the students in class 4/5-307 were born.

Icon	Label	Variables A
○		Months
1	1	November
2	2	May
3	3	June
4	4	December
5	5	February
6	6	August
7	7	October
8	8	June
9	9	June
10	10	March
11	11	September
12	12	May
13	13	November
14	14	May
15	15	June
16	16	April
17	17	February
18	18	June
19	19	November
20	20	October
21	21	December
22	22	March
23	23	September
24	24	November
25	25	May
26	26	August
27	27	September
28	28	December
29	29	July

Fig. 17.2. Students' database using
Data Explorer (Edwards 1998)

Class 4/5-307's Birthday Months

Months	Count
January	0
February	2
March	2
April	1
May	4
June	5
July	1
August	2
September	3
October	2
November	4
December	3

Fig. 17.3. Table of birthday data using
Data Explorer (Edwards 1998)

Once the displays are constructed, ask the students to work independently and write as much as they can about the information given in each type of display. After an appropriate amount of time (on the basis of the class's ability), ask the students to compare their written work and answer the questions for discussion.

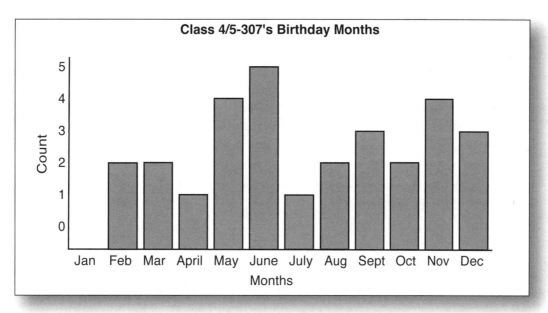

Fig. 17.4. Bar graph of birthday months constructed with Data Explorer (Edwards 1998)

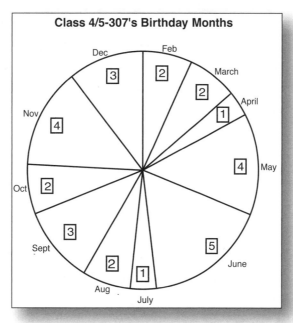

Fig 17.5. Circle graph of birthday months
constructed with Data Explorer (Edwards 1998)

Questions for discussion

1. "Which display (i.e., table, bar graph, or circle graph) did you find the easiest to understand? Why?" [RBY]

2. "How would you change any of the displays to make them easier to read and understand?" [RBY]

3. "How many students in our class were born in May? Which display did you use to answer this question?" [RD]

4. "What title did you give to the displays you prepared using the computer?" [RD or RBW]

5. "During which month(s) were the most students in our class born? Which display did you use to answer this question?" [RBW]

6. "During which month(s) were the fewest number of students in our class born? Which display did you use to answer this question?" [RBW]

7. "Do you think there are more students in our [grade or school] born in [insert month A] than in [insert month B]? Can you answer this question from any of the displays? Why? Why not? How can we find out the answer to this question?" [RBY] If this question is of interest to the students, it could lead to collecting information from other classes.

8. "For each display, ask a question that can be answered. How are your questions the same or different for each display?"

Writing and reading

For this activity, the writing occurs at the beginning, when the students are asked to write as much as they can about the information given in each type of display. The students share their written work to compare their ideas. For purposes of summarizing the similarities and differences of the three displays, include a culminating writing assignment.

Figure 17.6 shows the response of Garin, a fifth-grade student, to the directions in Appendix 15. Notice the three levels of comprehension reflected in the responses:

Reading the data: "There is [sic] 44 students in Mr. Kahn's class."

Reading between the data: "The most birthdays are being celebrated in June and November."

Reading beyond the data: "I liked display 3 because it shows fractions. It could teach kids about fractions because the more students there are the bigger the piece of the whole."

Garin   5-250

① There is 44 students in Mr. Kahn's class.
   The most birthdays are being celebrated in June and November.
   There is no birthdays in February.

② I liked display 3 because it shows fractions. It could teach kids about fractions because the more students there are the bigger the piece of the whole. Number 3 was easiest.

   I think all the displays are good.

Fig. 17.6. Garin's written work

Activity 18

Topic	A comparison of height and standing long jump distance
Graph form	Double bar graph
Graph title	Can You Jump Your Height?
Objectives	1. To determine one's height by measuring in customary units of measure
	2. To determine the distance one is able to jump (see fig. 18.1)
	3. To collect, organize, and interpret data
	4. To construct a double bar graph
	5. To compare the height and distance jumped for each member of a small group of students
	6. To use the computer as a graphing tool
	7. To answer comprehension questions the basis of information in a specific graph
	8. To express the meaning of a graph in prose
Vocabulary	Graph, double bar graph, height, standing long jump, compare
Materials	Approximately ten customary-unit tape measures attached in sets of two; a standing long jump mat (see fig. 18.1); 1/4" graph paper (Appendix 7) or 5-mm graph paper (Appendix 8), pencils, rulers, and data sheets (Appendix 13) for each student; computers and graphing software

Fig. 18.1. Photo of standing long jump mat

Procedure

Ask the students, "Do you think you can jump your height? One of the expectations of a common physical fitness test is that you be able to jump your height. Let's see whether you can do this." This activity should be fun; the results should not be used for a physical fitness grade.

Allow the students to work in small groups of four or five so that they can measure one another's height (see Activity 16) and distance jumped. Review how to read a tape measure and how to add on the amount from the second tape measure. Show the students how to measure height by placing a guide (such as a ruler) gently on top of the head of the student being measured (see fig. 16.1). Have each student record the heights of all the members of his or her group, using a data sheet similar to Appendix 13.

A standing long-jump mat is usually marked in feet. If a mat is not available, mark off a distance of seven feet on the floor in one-foot intervals. (Be advised that not using a mat may be dangerous, depending on the floor surface.) Ask each student to record the distances jumped for all the members of his or her group (see fig. 18.2).

Once the data have been collected, the students in each group discuss how to organize the bar graph. Since height is involved, suggest that the bars be vertical. Distribute graph paper and have the students draw a set of axes and label them. They will have to decide whether to record the

Fig. 18.2. Sixth grader jumping

results in inches, feet, or a combination of the two. The students should also decide what scale to use so that everyone's height and distance jumped can fit on the graph properly. Remind the students that equal spacing between bars and between the numerals listed along the vertical axis is necessary. Have them insert a title on the graph.

Using technology

A multiple bar graph option should be available on the software selected. After recording the heights and distances jumped for the members in their groups, the students should work at the computers in pairs to enter the data, construct the graph, and explore other software options (e.g., adjust the scaling and convert to different graph formats). Have them print out their graphs for comparison; the data and the graphs could be stored on disks for future reference and use.

Questions for discussion

1. "What is this graph about?" [RD or RBW]
2. "Who is taller, [insert name] or [insert name]?" [RBW]
3. "Who jumped farther, [insert name] or [insert name]?" [RBW]
4. "Who is the tallest?" [RBW]
5. "Who jumped the farthest?" [RBW]
6. "Who is able to jump at least his or her height?" [RBW]
7. "Why do you think that it is a measure of physical fitness to be able to jump your height?" [RBY]
8. "Think of a question you can ask about this graph."

Writing and reading

After the students have discussed the answers to the questions, ask them to write a description of what the graph means to them. Ask them to share their work by reading their descriptions aloud or by switching papers among groups. Allow group members to compare the graphs with the written descriptions of them. Allow the students to analyze one another's work, encouraging them to question and criticize constructively.

Activity 19

Topic	Height compared to foot length (or forearm length)
Graph form	Double bar graph
Graph title	How Does Your Height Compare with Your Foot Length (or Forearm Length)?
Objectives	1. To measure one's height and foot length (or forearm length) using metric units
	2. To collect, organize, and interpret data
	3. To construct a double bar graph
	4. To compare height and foot length of students in small groups of four or five
	5. To use the computer as a graphing tool
	6. To answer comprehension questions on the basis of information in a specific graph
	7. To express the meaning of a graph in prose
Vocabulary	Graph, double bar graph, height, foot length (or forearm length), metric units of measure, centimeter, meter
Materials	Chalkboard and chalk (document projector or overhead projector, transparencies, and markers); 1/4" graph paper (Appendix 7) or 5-mm graph paper (Appendix 8), pencils and rulers for each student; metric tape measures attached to wall; recording sheet (similar to Appendix 13); computers and graphing software

Procedure

Ask the students, "How do you think a person's height and foot length are related?" Allow them to discuss what they perceive this relationship to be.

Allow the students to work in groups of four or five to measure one another's height. Attach two metric tape measures end-to-end to the wall in several places around the classroom (see Activity 16). Show the students how to measure one another's height as described in Activity 18. Ask each student to record the heights of all the members of his or her group on a data sheet similar to Appendix 13.

Groups of students not at a height station should measure their foot length by taking off a shoe and standing on a metric tape measure or ruler. Students are to work in pairs so that the foot measure can be read properly. Explain how to place the foot on the tape measure or ruler and how to read it. Have each student record the foot lengths of all the members of his or her group.

Once the data have been collected, the students in each group should discuss how to organize the bar graph. Because height is involved, vertical bars are appropriate. Distribute graph paper and have the students draw a set of axes and label them. Allow the students to decide what scale to use so that everyone's height can fit on the graph properly. Remind the students that equal spacing between bars and between numerals listed along the vertical axis is necessary. Have them insert a title on the graph.

Using technology

See comments for Activity 18. Depending on students' interest, they may want to use the computer to make a scatter plot and explore the relationship between height and foot length (or forearm length).

Questions for discussion

1. "What is this graph about?" [RD or RBW]
2. "How tall is [insert name]?" [RD]
3. "How long is [insert name]'s foot?" [RD]
4. "Who is the tallest? Who is the shortest?" [RBW]
5. "Whose foot is the longest? Whose foot is the shortest?" [RBW]
6. "Who is taller, [insert name A] or [insert name B]? Whose foot is longer, [insert name A] or [insert name B]?" [RBW]
7. "From the data you have collected, what is the relationship between a person's height and his or her foot length? Do you think you have collected enough data to conclude that this is generally true? How can we be sure?" [RBW and RBY]
8. "Think of a question you can ask about this graph."

Writing and reading

After the students have discussed the answers to the questions, ask them to write a description of what the graph means to them. Ask them to share their work by switching papers, data sheets, and graphs among the groups. Allow students to analyze one another's work, encouraging them to question and criticize constructively.

Activity 20

Topic	Raisin experiment
Graph form	a. Bar graph
	b. Line plot
	c. Stem-and-leaf plot
	d. Box plot
	e. Double bar graph
	f. Back-to-back stem-and-leaf plot
	g. Multiple box plot
Graph title	How Many Raisins Are in a 1/2-oz. Raisin Box?
Objectives	1. To estimate the number of raisins in a 1/2-oz. raisin box

2. To count the raisins in a 1/2-oz. raisin box
3. To collect, organize, and interpret data
4. To categorize estimates and counts according to similar responses
5. To construct a bar graph, line plot, stem-and-leaf plot, box plot, double bar graph, back-to-back stem-and-leaf plot, and multiple box plot
6. To use the computer as a graphing tool
7. To use the graphing calculator as a tool
8. To identify and discuss the advantages and disadvantages of using different graph forms to display data
9. To answer comprehension questions on the basis of information in a specific graph

10. To express the meaning of a graph in prose

Vocabulary Graph, bar graph, double bar graph, line plot, stem-and-leaf plot, back-to-back stem-and-leaf plot, box plot, multiple box plot, estimate, range, number line, median, lower quartile, upper quartile, outlier, mode

Materials Chalkboard and chalk (document projector or overhead projector, transparencies, and markers); 5-mm graph paper (Appendix 8), lined paper, data recording sheet (Appendix 17), napkins, 1/2-oz. boxes of raisins (all of the same brand), pencils, rulers, two different colors of Post-It notes (3" × 3") for each student; computers and graphing software; a graphing calculator for each student

Procedure

Hold up a 1/2-oz. box of raisins and ask students, "How many raisins do you think are in this box?" Write a few estimates on the chalkboard (document projector or on a transparency).

Distribute boxes of raisins and Post-It notes of one color to all students. Ask them not to open the boxes yet. They must estimate how many raisins are in their boxes and write their estimates (making them large and dark) on the Post-It notes. If you want to keep a record of individual estimates after the actual counting is done, have the students write their initials in the corner of the notes. Have the students post their estimates on the chalkboard, and discuss with them how to organize the estimates by posting the same estimates in columns. The students should record the data on a sheet similar to Appendix 17 (see fig. 20.1). (Collect the boxes of raisins if the counting is to be done in another lesson. Be sure to save the Post-It notes of the estimates to compare with actual counts in subsequent lessons.)

NAME OF STUDENT	ESTIMATE	ACTUAL COUNT
Chris	25	
Reagan	20	
Amie	25	
Ann	24	
Jamie	25	

Fig. 20.1. Students' entries on a data sheet

a. *Bar graph:* Distribute graph paper. Have the students construct a set of axes and label them. It is suggested that the bars be arranged vertically so that the bar graph can be compared with the line plot in the next section. Discuss the range of the estimates and how to list them along the horizontal axis. Depending on the range of estimates, discuss the possibility of grouping the data (e.g., estimates between 10 and 14; 15 and 19; etc.). Be sure that students insert a graph title (see fig. 20.2). Discuss the answers to questions 1–3 and 13 below.

b. *Line plot:* Have the students draw a number line identifying the lowest and highest estimates. Then have them construct a line plot, using the information from the Post-It notes on the chalkboard (or wall or chart). Be sure that the spacing between numerals is equal (see fig. 20.3).

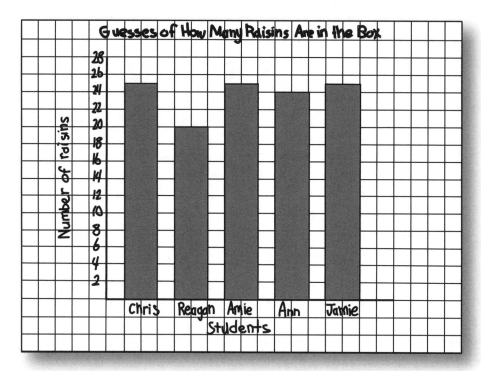

Fig. 20.2. A sixth grader's graph of estimates

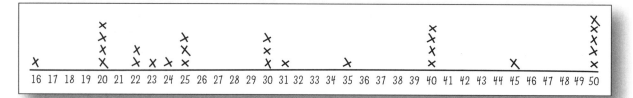

Fig. 20.3. Sixth graders' example of a line plot for estimates of raisins in a box

Ask the students to compare the information displayed in the line plot with the information displayed in the bar graph (with ungrouped or grouped data). Discuss the answers to questions 1–5 and 13.

c. *Stem-and-leaf plot:* Since most of the estimates will probably be two-digit numbers, have the students arrange the Post-It notes on the board in rows according to the tens digit (i.e., all single-digit estimates in one row, all estimates in the tens in the second row, etc.). Have students convert this listing to a stem-and-leaf plot (see fig. 20.4). Identify the lowest estimate, the highest estimate, the median, and the mode. (Be sure to review the meaning of median and mode.)

TENS	ONES
1	6
2	00002234555
3	00015
4	00005
5	00000

1 | 6 means 16

Fig. 20.4. Sixth graders' example of a stem-and-leaf plot
for estimates of raisins in a box

Have the students rotate their stem-and-leaf plots 90 degrees counterclockwise and compare them with the bar graph (ungrouped or grouped data) and the line plot. Allow the students to discuss the similarities and differences. Discuss the answers to questions 1–7, 12, and 13.

d. *Box plot:* Depending on the students' ability, have them construct a box plot of the estimates by identifying the lowest estimate, the highest estimate, the median, the lower quartile, and the upper quartile (see fig. 20.5). Discuss the answers to questions 1–3, 8, 9, 12, and 13.

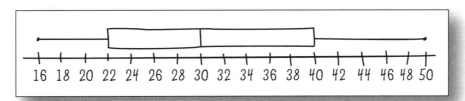

Fig. 20.5. Example of a box plot using data from estimates of raisins in a box

After the students have analyzed the data from the estimates, they are ready to count the number of raisins in their boxes. Distribute napkins, boxes of raisins (if this is a continuation lesson), and Post-It notes of a different color; have the students write their actual counts (making them large and dark) on the notes. Have the students write their initials in the corner of the Post-It note if you want to keep a record of individuals' actual counts to compare with their estimates (see question 3a below). Have the students arrange the actual counts on the chalkboard similarly to the way in which the estimates were arranged. (If this is a continuation lesson, reattach the Post-I® notes of the estimates.) Have the students discuss what they are observing about the actual counts in comparison to the estimates.

Using the data from the actual counts, the students can construct a bar graph, a line plot, a stem-and-leaf plot, and a box plot. Depending on the students' ability, they may go on to comparing the estimates and the actual counts formally by constructing a double bar graph, a back-to-back stem-and-leaf plot, or a multiple box plot.

e. *Double bar graph:* Have the students group the data from the estimates and the data from the actual counts so that it will be possible to construct a double bar graph. Allow them to decide whether the bars should be vertical or horizontal. Be sure they label the axes, insert a title, and include a key to represent estimates and actual counts. A sample double bar graph is shown in figure 20.6. Discuss the answers to questions 1, 2b, 3b, 11, and 13.

f. *Back-to-back stem-and-leaf plot:* Using the stem-and-leaf plot constructed for the estimates, allow the students to attach a column to the left to represent the units digits of the actual counts (see fig. 20.7). Have them identify the lowest actual count, the highest actual count, and the median. Using this back-to-back plot, ask them to compare the estimates to the actual counts. Also, have them compare the information displayed in the back-to-back stem-and-leaf plot with the information compared in the double bar graph. Discuss answers to questions 1, 2b, 3b, and 10–13.

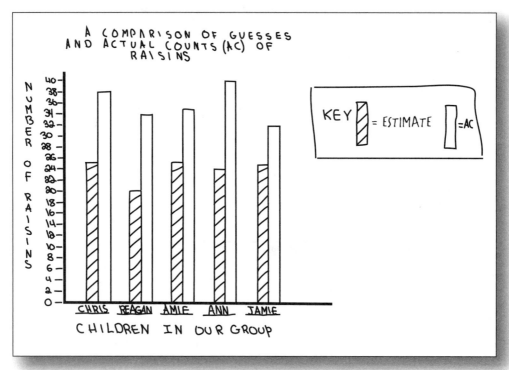

Fig. 20.6. A sixth grader's double bar graph

Actual Counts		Estimates
ONES	TENS	ONES
	1	6
9	2	00002234555
99999888855422221	3	00015
11100000	4	00005
	5	00000

9 | 1 | 6 means 19 actual count
                        16 estimate

Fig. 20.7. Sample of back-to-back stem-and-leaf plot using data of actual counts and estimates of raisins in a box

g. *Multiple box plot:* Using one axis to indicate lows, highs, medians, and lower and upper quartiles, have the students construct a box plot for the estimates and the actual counts (see fig. 20.8). Have them compare the results of the estimates with the actual counts. Also, ask them to compare the information displayed in the double bar graph, the back-to-back stem-and-leaf plot, and the multiple box plot. Discuss the answers to questions 1–3, 2b, 3b, 9, and 11–13.

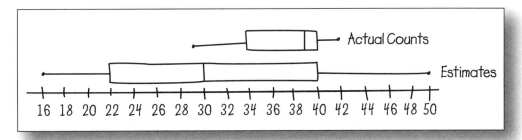

Fig. 20.8. Sample of a multiple box plot using data of actual count and estimates
of raisins in a box

Using technology

Computer graphing software that has the capability of constructing stem-and-leaf plots and box plots, as well as bar graphs and double bar graphs, is desirable for this activity.

After the students have had experience constructing the graphs and plots by hand, give them the opportunity to explore the graphing and plotting options of various pieces of software. Perhaps some groups of students could examine possibilities for graphing the data using bar graphs or double bar graphs, and others could use software to apply plotting techniques. The groups then exchange the software to complete their analyses.

Ask students to print out the graphs and plots to examine and discuss similarities and differences. Students could save the data, graphs, and plots on a disk for future reference and use.

If students have access to graphing calculators, they may want to analyze the raisin data making use of this technology. The features of the graphing calculator that they will use will be "STAT," "WINDOW," "STAT PLOT," and "GRAPH." (Consult the calculator manual to decide which features you want the students to use and teach them how to use the calculators appropriately.) Using the "STAT LIST EDITOR," students should construct two lists—one for the estimates (e.g., "L1") and one for the actual counts (e.g., "L2"). See figure 20.9 for a partial viewing of the two lists in a screen capture from TI-84 Plus Silver Edition.

Fig. 20.9. Screen capture of partial listing and
estimates of the number of raisins in a box

Once the lists are complete, the students should select the "WINDOW" to adjust the display screen so that there is space to display two box plots. See figure 20.10 for a TI-84 Plus Silver Edition screen capture of an example of a window for displaying the two box plots.

Fig. 20.10. Screen capture of the WINDOW screen to display
box plots of the raisin data on the TI-84 Plus Silver Edition

The students will need to use this "STAT PLOT" feature to gain access to "Plot 1" and "Plot 2." Both plots need to be "on," the box plot icon needs to be highlighted, the "Xlist" for Plot 1 needs to be "L1," and the "Xlist" for Plot 2 needs to be "L2." Once all the settings are complete, students should select "GRAPH" and view the two box plots. See fig. 20.11 for a screen capture of the box plots on the TI-84 Plus Silver Edition.

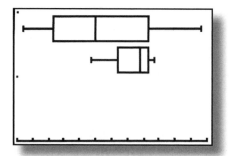

Fig. 20.11. Screen capture of box plots of the raisin data
using a TI-84 Plus Silver Edition

Using the TRACE feature, the students can read the five critical points (i.e., lowest datum, first quartile, median, third quartile, and the highest datum) directly from the plots. "P1:L1" refers to Plot 1 for List 1, and "P2:L2" refers to Plot 2 for List 2. The box plot of L1, that is, the estimates of the number of raisins in 1/2-oz. boxes, is the top plot, representing the greater variability. The lower plot represents the actual count, illustrating that the data are more concentrated and have less variability.

Questions for discussion

1. "What is this graph or plot about?" [RD or RBW]

2. a. "Which estimate occurred most frequently [i.e., the mode]? Which estimate occurred the fewest number of times?" [RBW]

   b. "Which actual count occurred most frequently [i.e., the mode]? Which actual count occurred the fewest number of times?" [RBW]

3. a. "Which estimate is the lowest? The highest? What is the range of the estimates?" [RBW]

   b. "Which actual count is the lowest? The highest? What is the range of the actual counts? Whose estimate was the closest to his or her actual count?" [RBW]

4. "Using the line plot or stem-and-leaf plot, how can you find the median? What is the median of the estimates?" [RBY]

5. "What is similar about a bar graph and a line plot? What is different? Which do you prefer? Why?" [RBY]

6. "What is similar about a stem-and-leaf plot and a bar graph? What is different? Which do you prefer? Why?" [RBY]

7. "What is different about a line plot and a stem-and-leaf plot? What is the same? Which do you prefer? Why?" [RBY]

8. "What is different about a bar graph and a box plot? Which do you prefer? Why?" [RBY]

9. "What information can you obtain from a stem-and-leaf plot that you cannot obtain from a box plot? When do you think it would be more appropriate to use a box plot than a stem-and-leaf plot?" [RBY]

10. "What is the same [or different] about a double bar graph and a back-to-back stem-and-leaf plot? Which do you prefer? Why?" [RBY]

11. "What brand of raisins did we use? How do you think another brand would compare?" [RBY]

12. "How can we determine which is the best graph form to use to display our data?" [RBY]

13. "Think of a question you can ask about the graphs and plots."

Writing and reading

After the students have constructed and discussed different graphs and plots, ask them to write about how they constructed the graph and what it means to them. Ask them to share their work by switching papers, data sheets, and graphs and plots. Allow the students to analyze one another's work, encouraging them to question and criticize constructively.

Note: This is not meant to be an introduction to line plots, stem-and-leaf plots, box plots, or the use of the graphing calculator. The concepts involved should be developed prior to doing the respective parts of this activity (see Landwehr and Watkins 1986). Also, all the parts of this activity are not meant to be done in one lesson.

There could be several variations or extensions of this activity. The students could work in small groups and compare estimates and actual counts among themselves using bar and double bar graphs only, or they could use different brands of 1/2-oz. boxes of raisins. Both estimates and actual counts of each could be compared and discussed, thus raising issues of packaging, weight, size of raisins, and so on.

Some ideas for this activity were taken from Helene Silverman (1987, 1988). Similar ideas can be found for estimating the number of M&M's in bags of a certain size and analyzing the actual colors after counting (see Corwin and Friel 1988).

Activity 21

Topic	Height of a plant over a period of time
Graph form	Line graph
Graph title	Height of My Plant from [Insert Date] to [Insert Date]
Objectives	1. To plant a bean (or seed) and record the height of the resulting plant over a set period of time
	2. To collect, organize, and interpret data

3. To determine the difference between data displayed in a bar graph and data displayed in a line graph

4 To construct a line graph

5. To use the computer as a graphing tool

6. To answer comprehension questions on the basis of information in a specific graph

7. To express the meaning of a graph in prose

Vocabulary   Graph, line graph, height, a period of time, lima bean, seed, plant, soil, millimeters, centimeters

Materials   Chalkboard and chalk (document projector or overhead projector, transparencies, and markers); small water pitchers; adhesive labels, old newspapers, plastic or Styrofoam cups, soil, lima beans or other seeds, centimeter rulers, 5-mm graph paper (Appendix 8), data recording sheet (Appendix 18), and pencils for each student; computers and graphing software

Procedure

Ask the students, "If we plant a lima bean (or other seed), how tall do you think the plant will grow in three weeks?" Write the students' responses on the chalkboard. Say to them, "So, all we have to do is plant this bean [or seed], and it will grow? What does a plant need to grow? [water, light, air, and nutrients from the soil]. How must we care for our planted beans [or seeds] to help them grow into plants?"

Distribute adhesive labels so that the students can write their names on them to label their plants. Then distribute newspapers to protect desks, plastic or Styrofoam cups with soil, and two beans or other seeds per student. Have the students insert a finger about 2.5 cm into the soil, place a lima bean or other seed into the hole, and cover it. They should leave some space and plant the other seed similarly (see fig. 21.1). After covering the seeds, have them moisten the soil. (Avoid applying too much water, which would drown the seeds.) Place all the cups on a windowsill so that they can be exposed to sunlight, and assign students the responsibility for watering the plants daily.

Fig. 21.1. Students planting seeds

Once the plants begin to break through the soil, have the students measure the height in millimeters (see fig. 21.2) and record it on a data collection sheet (see fig. 21.3).

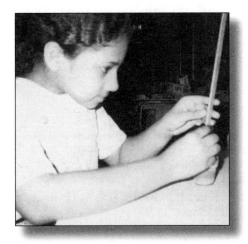

Fig. 21.2. Measuring the height of a plant

Height over Time
Data Collection

Name _Katherine_   Date _6/5_

School _P.S 11_   Class _4-407_

Due: _6/12_

Keep a record of your height (or the height of a plant) for a given period of time.

Date	Height (in cm)
6/5	10.5 cm
6/9	12 cm
6/10	15 cm
6/11	17.5 cm
6/12	17, 5 cm

Construct a line graph using the data above. Write a story about your graph.

Fig. 21.3. Katherine's data for the height of a plant for one week

After collecting data for a given period of time (e.g., one, two, or three weeks), the students should construct a line graph. Distribute graph paper and have them draw a set of axes. Indicate the dates (equally spaced) along the horizontal axis and the height (beginning at 0 where the two axes intersect) on the vertical axis. Review the meaning of the lines and boxes. Be sure that the numerals written along the vertical axis are equally spaced. Have the students label the axes, plot the points, connect the points, and insert a title on the graph. See figure 21.4.

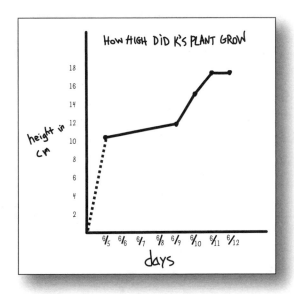

Fig. 21.4. A line graph of Katherine's data

Using technology

Select graphing software that can be used to construct a line graph as well as other graph forms. Once students are familiar with the software, allow them to work at computers in pairs to enter the data, construct the graphs, and compare the results of displaying the data in different graph forms. Also have them experiment with different scales. To facilitate the discussion of comparisons, ask the students to print out the graphs. The data and the graphs could be saved on a disk for future reference and use.

Questions for discussion

1. "What is this graph about?" [RD or RBW]
2. "How tall was the plant on [insert a date]?" [RD]
3. "On which date was the plant about half its size in comparison to the height recorded on the last date?" [RBW]
4. "Did all the beans [or seeds] grow? Why? Why not?" [RBY]
5. "What were the conditions that contributed to the growth of your plant? Could you have improved any of the conditions so that the plant would have grown faster or taller? How could we investigate this?" [RBY]
6. "Why did we use a line graph and not a bar graph to record the height of our plants?" [RBY]
7. "Think of a question that you can ask about your graph."

Writing and reading

After the students have discussed the answers to the questions, ask them to write a description of the activity and the graph, telling how they conducted the experiment and what the graph means to them. Writing about the activity can also be in the form of an ongoing log or journal. Ask them to share their work by switching papers, data sheets, and graphs. Allow students to analyze one another's work, encouraging them to question and criticize constructively.

Activity 22

Topic	Height (or weight) over time (Note: For some students, height or weight may be a sensitive issue. Also, the data may have to be collected for four or five months before the graph can be completed.)
Graph form	Line graph
Graph title	How Much I Have Grown (or Gained) from [Insert a Date] to [Insert a Date]
Objectives	1. To measure one's height or weight in metric units
	2. To collect, organize, and interpret data
	3. To construct a line graph
	4. To determine the difference between data displayed in a bar graph and data displayed in a line graph
	5. To use the computer as a graphing tool
	6. To answer comprehension questions on the basis of information in a specific graph

> 7. To express the meaning of a graph in prose
>
> *Vocabulary*   Graph, line graph, height, weight, metric units, centimeters, meters (or kilograms)
>
> *Materials*   Metric tape measures attached to wall (see Activity 16) or a metric scale; data collection sheets (Appendix 18), 5-mm graph paper (Appendix 8), rulers, and pencils for each student; computers and graphing software

Procedure

Ask the students, "How much do you think you will grow during the next four or five months? How can we find out if your prediction comes true?" [Keep a record of height.] This could be a continuation of Activity 16, where students were measured for constructing a small-group bar graph.

Allow the students to work in groups of four or five to measure one another's height. Have each student record the date and his or her height on a data collection sheet. This should be done consistently once or twice a month for four or five months (e.g., the first and third Mondays of the month).

Once the data (i.e., approximately eight measures of height for each individual) have been collected, discuss how to organize and display the data. Distribute graph paper and have the students draw a set of axes. They should space the dates equally along the horizontal axis and list numerals for height (in centimeters) along the vertical axis. Since the vertical axis is to begin with zero, discuss appropriate scales and spacing of the numerals so that the highest measure will fit on the graph. Be sure to review the meaning of the boxes and lines and how to plot a point located by the intersection of two grid lines. The axes should be labeled and a title inserted on the graph.

Using technology

See Activity 21.

Questions for discussion

1. "What is this graph about?" [RD or RBW]
2. "How tall were you on [insert a date]?" [RD]
3. "When were you [insert a measure] centimeters tall?" [RD]
4. "How much did you grow between [insert a date] and [insert a date]?" [RBW]
5. "How old were you at the time of the first measure? At the time of the last measure? Can this information be found directly from the graph? What must be done?" [RBY]
6. "How much do you think you will grow by the time you reach the [insert a grade level] grade?" [RBY]
7. "In Activity 16, we constructed a bar graph to represent heights. In this activity, we constructed a line graph. What is different about the data collected in this activity that requires a different type of graph?" [RBY]
8. "During the time you have been recording your height, how frequently did you have to buy new clothes because you outgrew your old ones, not because you wore them out?" [RBY]
9. "Think of a question you can ask about the graph."

Writing and reading

After the students have discussed the answers to the questions, ask them to write a description of the activity and the graph. Writing about the activity can be in the form of an ongoing log or journal. Ask them to share their work by switching papers, data sheets, and graphs. Allow the students to analyze one another's work, encouraging them to question and criticize constructively.

Activity 23

Topic	Daily temperature
Graph forms	a. Line graph
	b. Multiple line graph
Graph title	a. What Are Typical Morning (or Afternoon) Temperatures over a Seven-Day Period?
	b. How Does the Temperature Vary from Morning to Afternoon over the course of Five Days?
Objectives	1. To determine the outdoor temperature in the morning and the afternoon by reading a thermometer
	2. To collect, organize, and interpret data
	3. To construct a line graph and a multiple line graph
	4. To determine the difference between data displayed in a line graph and data displayed in a bar graph
	5. To compare morning and afternoon temperatures using a multiple line graph
	6. To use the computer as a graphing tool
	7. To answer comprehension questions on the basis of information in a specific graph
	8. To express the meaning of a graph in prose
Vocabulary	Graph, line graph, multiple line graph, key or legend, temperature, thermometer, Celsius, Fahrenheit, degrees
Materials	Overhead projector (or document projector), transparency of a thermometer scale (either Celsius or Fahrenheit); an outdoor Celsius or Fahrenheit thermometer; a data collection sheet (Appendix 19); 5-mm graph paper (Appendix 8), colored pencils, and pencils for each student; computers and graphing software

Procedure

Ask the students, "How does the outside temperature vary from the morning to the afternoon? From the morning to the evening? How can we be sure?" [Keep a record over the course of several days.]

Distribute data collection sheets to the students. Using an overhead transparency of a thermometer scale (either Celsius or Fahrenheit, to correspond with the scale to be used), review how to read a thermometer. Attach an outdoor thermometer to a window so it is clearly visible to the students. For five or more consecutive days, have the students record the temperature at the same time (e.g., 8:45 a.m. and 2:30 p.m.).

a. *Line graph:* Depending on the students' ability, graph morning and afternoon temperatures separately, or collect data only for morning or afternoon; adjust the graph title accordingly. Distribute graph paper and have the students construct a set of axes. Along the horizontal axis, list the days or dates, equally spaced, that the temperature was taken. Along the vertical axis, list the temperatures. Because zero is placed where the two axes intersect, any temperature below zero will require the extension of the vertical axis below the horizontal axis (see fig. 23.1). Be sure that the numerals along the vertical axis are equally spaced. Have students label the axes and insert an appropriate title. Discuss the answers to questions 1, 2, and 5–7 below.

b. *Multiple line graph:* Distribute graph paper and have the students construct a set of axes. After setting up the axes and labeling them, have the students select two colored pencils—one color to represent the morning temperature and one for the afternoon temperature. If colored pencils are unavailable, a dotted or dashed line can represent morning temperature, for example, with a solid line for the afternoon temperature. Be sure to include a key and an appropriate title (see fig. 23.2). Discuss the answers to all the questions below.

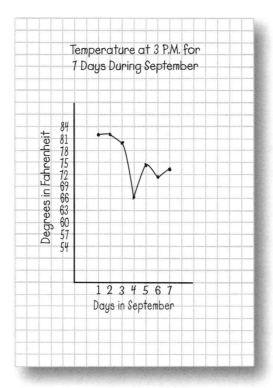

Fig. 23.1. Line graphs (Source: The National Weather Service)

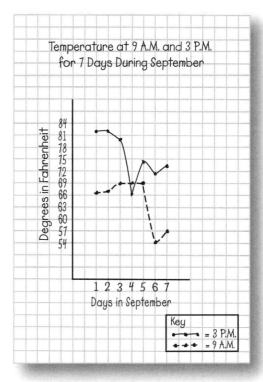

Fig. 23.2. Multiple line graph (Source: The National Weather Service)

Using technology

Select graphing software that can be used to construct line graphs and multiple line graphs. Once the students are familiar with the software, allow them to work at computers in pairs to enter the data, construct the graphs, explore alternative graph forms, and adjust the scale. Allow sufficient time for students to discuss their results. If a printer is available, the graphs could be printed out to facilitate making comparisons. Also, the data and the graphs could be saved on a disk for future reference and use.

Questions for discussion

1. "What is this graph about?" [RD or RBW]

2. "What was the temperature on [insert a date] in the morning? In the afternoon?" [RD]

3. "On [insert a date], when was the temperature higher, in the morning or the afternoon?" [RBW]

4. "Do you notice any temperature patterns from the morning to the afternoon during the five days? If so, what do you notice? Do you think this happens most of the time? Why do you think so? How can we be sure?" [One possible answer: Extend the experiment for seven or ten days.] [RBW/RBY]

5. "Why do you think people pay attention to the temperature and the weather? On whom do they usually rely for weather forecasts? Do you think the weather forecast is always accurate? Why? Why not? About how many times out of ten do you think the weather prediction is accurate? How can we be sure?" [One possible answer: Conduct an experiment to compare weather predictions with actual weather over a period of about ten days.] [RBY]

6. "Why were the data collected in this activity displayed in a line graph [or multiple line graph] and not a bar graph [or double bar graph]?" [RBY]

7. "Think of a question you can ask about this graph."

Writing and reading

After the students have discussed the answers to the questions, ask them to write a description of the activity and the graph. Have them share their work by switching papers, data sheets, and graphs. Allow students to analyze one another's work, encouraging them to question and criticize constructively.

Activity 24

Topic	Times of sunrise and sunset
Graph forms	a. Line graph
	b. Multiple line graph
Graph title	a. What Time Does the Sun Rise (or set) during a Thirteen-Day Period?
	b. How Does the Amount of Daylight Vary over a Thirteen-Day Period?
Objectives	1. To determine the times of sunrise and sunset by checking the times in the newspaper
	2. To collect, organize, and interpret data
	3. To construct a line graph and a multiple line graph
	4. To determine the difference between the data displayed in a line graph and the data displayed in a bar graph, picture graph, or circle graph
	5. To compare times of sunrise and sunset over a thirteen-day period using a multiple line graph
	6. To answer comprehension questions on the basis of information in a specific graph

7. To complete a graph of the average time of sunset for one year, given a graph of the average time of sunset for six months

8. To express the meaning of a graph in prose

Vocabulary   Graph, line graph, multiple line graph, key, legend, sunrise, sunset, time, seasons, daylight

Materials   Data collection sheets (Appendix 20), 5-mm graph paper (Appendix 8), pencils, and colored pencils for each student; Graph Completion Task (Appendix 21)

Procedure

Ask the students, "What time did the sun rise this morning? What time will the sun set this evening? For whom might these be important questions? Why would these be important questions? How can we find out the times of sunrise and sunset?"

Distribute data collection sheets to the students and have them obtain the times of sunrise or sunset over a thirteen-day period by looking in the daily newspaper.

a. *Line graph:* Have the students construct a line graph of either sunrise or sunset or make two separate line graphs of each. Distribute graph paper and have the students construct a set of axes. Have the students list the dates so that they are equally spaced along the horizontal axis and list the times along the vertical axis. One approach would be to use the point of intersection of the two axes to represent midnight (or the start of the new day). Discuss ways of labeling the times, since the sunsets during a thirteen-day period will have a daily time difference of approximately one minute (see fig. 24.1). Discuss the answers to questions 1–3 and 5.

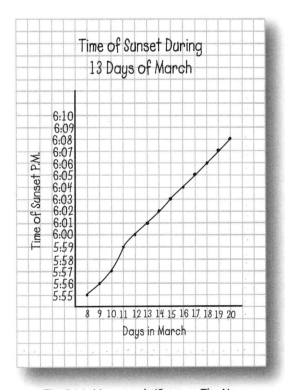

Fig. 24.1. Line graph (Source: *The New York Times*)

Be sure that the axes are labeled and an appropriate title is inserted. Assist students if they need help plotting the points. They may need to be reminded about what the boxes and lines in the graph represent and how to plot a point, which is located by the intersection of two grid lines.

b. *Multiple line graph:* Distribute graph paper and have the students construct a set of axes. Along the horizontal axis, they should list the dates, equally spaced. Since the times of both sunrise and sunset will be listed along the vertical axis, discuss the possibilities for doing this properly and accurately. Students may have to make a "break" in the graph in order to make all the data fit (see fig. 24.2). Discuss the answers to all the questions below.

Ask students to select two colors—one to represent times of sunrise, the other for times of sunset. Be sure that a key is included. If colored pencils are unavailable, use a solid line for the times of sunrise, for example, and a dotted or dashed line for the times of sunset. Axes are to be labeled and an appropriate title inserted.

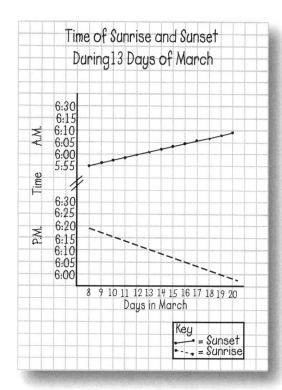

Fig. 24.2. Multiple line graph (Source: *The New York Times*)

As an example of an alternative assessment task, use Appendix 21, "A Graph Completion Task," which asks students to read beyond the data in order to complete the average time of sunset from January to June. The completed graph will represent the average time of sunset in the eastern time zone at the latitude of New York (see fig. 24.3 for an example of four students' work).

Students may want to investigate how the graph may be different for various cities in Australia, Russia, or other parts of the world. This may also lead to an extranet experience. "An extranet has two principal features, namely, the ability to (1) display Web pages, which may be password protected, to authorized individuals anywhere in the world, and (2) automatically accept, process, and distribute data using Web-based forms" (English and Cudmore 2000, p. 83).

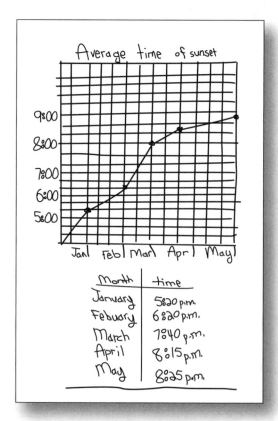

Fig. 24.3. Solution to graph task created by Chuck, Dennis, Garin, and Razzie

Questions for discussion

1. "What is this graph about?" [RD or RBW]

2. "At what time did the sun rise [or set] on [insert date]?" [RD]

3. "On what day did the sun rise [or set] at [insert time]?" [RD]

4. "During which seasons are we recording the times that the sun rises [or sets]? As the days progress, what do you notice about the time of sunrise [or sunset]? What is happening to the number of daylight hours during this season? How would this be the same [or different] during [insert another season]?" [RBY]

5. "What are the characteristics of the data collected in this activity that require them to be displayed in a line graph and not a bar graph, picture graph, or circle graph?" [RBY]

6. "How might the times of sunrise [or sunset] for [insert a particular month] differ in, for example, Australia? Why do you think this is so? How could we find out?" [RBY]

7. "Think of a question you can ask about this graph."

Writing and reading

After the students have discussed the answers to the questions, ask them to write a description of the activity and the graph. Writing about the activity can be an ongoing log or journal. Ask the students to share their work by switching papers, data sheets, and graphs. Allow them to analyze one another's work, encouraging them to question and criticize constructively.

 # Activity 25

Topic	Social justice: Poor vs. rich students' access to formal education in selected foreign countries
Graph form	Double bar graph
Graph title	Average Years of Education of the Poorest and Richest 17-to-22-Year-Olds, in Selected Countries, 2005
Objectives	1. To locate selected countries on a map of the world
	2. To construct a double bar graph using data from UNESCO
	3. To compare average years of education of poorest and richest 17-to-22 year olds in selected foreign countries
	4. To use the computer as a graphing tool
	5. To answer comprehension questions on the basis of information in a specific graph
	6. To write a letter of opinion using data in a double bar graph
Vocabulary	Graph, double bar graph, equitable formal education, social justice
Materials	Chalkboard and chalk, document projector, or overhead projector, transparencies, and markers; 1/4" graph paper (Appendix 7) or 5-mm graph paper (Appendix 8), pencils and rulers for each student; map of the world to locate foreign countries; activity sheet (Appendix 22); computers and graphing software

Procedure

Ask the students, "Do you think all children around the world have equal opportunity?" "How do you think opportunities for formal education compare for students in different countries around the world?" As students respond, post their ideas on the board, overhead projector, or document projector.

Follow up with, "How do you think children from poor families and rich families compare in the number of years they go to school in different countries?" As students respond, post their ideas on the board, overhead projector, or document projector.

Have a list of the countries posted (i.e., Bangladesh, Burkina Faso, Ghana, Guatemala, India, Mozambique, Nicaragua, Nigeria, Peru, Philippines, Tanzania). Have students locate the countries on a world map, identifying the continents (i.e., Europe, Asia, Africa, North America, South America, Antarctica, and Australia) where the countries are located (see fig. 25.1).

Fig. 25.1. Two seventh graders reading a map of the world

Distribute the activity sheet so students can fill in the continents and begin to work on constructing a double bar graph using the computer and answering the questions.

Using technology

A multiple bar graph option should be available on the software selected. If available, Microsoft Excel can be used to enter the data and to construct a double bar graph (see Fig. 25.2). Have students work in pairs to enter the data into the computer (see Fig. 25.3), construct the graph, and answer the questions on the accompanying activity sheet (Appendix 22). On completion, review the answers in a whole-class discussion or with small groups of students.

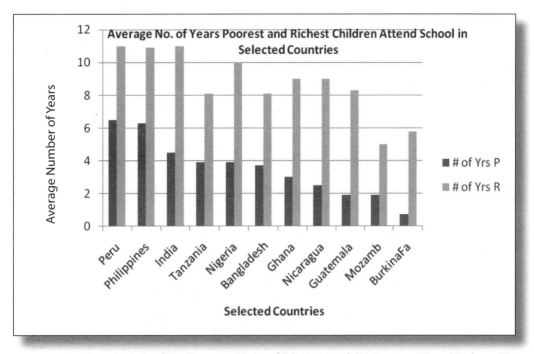

Fig. 25.2. Students' graph of the "Average Years of Education of the Poorest (P) and Richest (R) 17-to-22-Year Olds in Selected Countries, 2005," created with Microsoft Excel

Fig. 25.3. Three students analyzing the graph they constructed using the computer

Questions for discussion

1. "What is this graph about?" [RD or RBW]
2. "Why do you think this graph is called a 'double bar graph'? What are the characteristics of a double bar graph?" [RBY]
3. "What is the average number of years that the poorest students in Ghana spend in school?" [RD]

4. "What is the average number of years that the richest students in Peru spend in school?" [RD]

5. "In which continent is each country located?" [map reading, RD]

6. " What do the data tell us about children's opportunity for education and poverty?" [RBW]

7. "In which country is the average number of years spent in school the least for the poorest students?" [RBW]

8. "In which country is the average number of years spent in school the greatest for the richest students?" [RBW]

9. "Of the countries listed, on which continent, on average, do the poorest students spend the least amount of time in school?" [RBY]

10. "Of the countries listed, on which continent, on average, do the richest students spend the greatest amount of time in school?" [RBY]

11. "What are the advantages of reading data in a double bar graph? What are the advantages of reading data in a table?" [RBY]

12. "Why do you think there are differences between the average number of years that the poorest and richest students spend in school in each of the countries listed in the table? How do you think public education in the United States attempts to provide equal opportunities among the poorest and richest students? How can we determine whether public education is equitable among poor and rich students?" [RBY]

13. "Using these data, what recommendations can we make? What additional information would be relevant?"

14. "Think of a question you can ask about this graph."

Writing and reading

After the students have discussed the answers to the questions, ask them to write a letter to UNESCO, making recommendations supporting equal opportunity for education (see fig. 25.4 for a letter written by a seventh grader). Ask them to share their work by switching papers, data sheets, graphs, and drafts of letters among the groups. Allow students to analyze one another's work, encouraging them to question and criticize constructively.

```
Dear UNESCO:

    After reading the data for the average number of years
spent in school for both rich and poor kids, I was shocked to see
how in some countries, they do not afford/give a lot of education
to the poor kids, but they give a lot to the rich. I think both
the rich and poor kids deserve equal education at an affordable
price. These data show how some countries aren't very fair with
some families' economic life styles. All countries need equality.
What can we do about this? How can I help?

    Thank you.

                                              Sincerely,
                                              Rani
```

Fig. 25.4. Letter to UNESCO written by a seventh grader

Activity 26

Topic:	The extent to which parents and their children agree and disagree
Graph form	Double bar graph
Graph title	How Much Do We Agree and Disagree?
Objective;	1. To develop a survey
	2. To conduct a survey
	3. To report the results of a survey in a double bar graph
	4. To interpret and analyze the results of a survey
	5. To compare the results of a class survey with the results of a professionally conducted survey
	6. To answer comprehension questions on the basis of the results of a class survey and a professionally conducted survey
	7. To write a letter reacting to a professionally conducted survey
Vocabulary	Survey, poll, respondents, questionnaire, double bar graph
Materials	Chalkboard and chalk, overhead projector transparencies and markers, or document projector; Appendix 23, pencils or pens, and lined writing paper for each student; word processing software, data analysis software, computer, printer

Procedure

Day 1: Ask students, "What are some things that you and your parents agree on?" "What are some things you disagree on?" Enlist students to record what their peers say on the board, on the overhead projector, or on the document projector. Some seventh graders admitted that they and their parents disagree about the music they like, the TV shows they like to watch, and the food they like to eat, and agree about the importance of an education and keeping safe. After discussing such disagreements and agreements, ask, "How can we be sure about these agreements and disagreements?" Elicit the need to ask their parents by conducting a survey. Ask, "What is a survey?" "How are surveys created?"

Duplicate Appendix 23 and cut the sheet in half. Distribute one half to part of the class and the other half to the other part of the class. Ask students to complete the short survey and compile the results. Bring the class together to discuss the advantages and disadvantages of the two types of surveys and agree on the type of survey to construct for the class project. Should the questions be open-ended? If so, why? Should the respondents have choices from which to select the answer? If so, why?

Using the ideas they discussed at the beginning of the lesson, have students organize the responses to the things they and their parents agree upon, posted according to appropriate categories: types of music (e.g., rock, country, rhythm and blues rap or hip hop, classical, jazz, salsa or Spanish rock), TV shows, types of food to eat, and so on. Assign each category to a small group of students to write questions to pose in a survey. Have students share their questions with the class. Conduct a discussion to help students critique the questions for clarity. Have students prepare the survey for distribution.

Homework: Have students conduct the survey of their parents and siblings and bring in the results the next day for compilation and discussion.

Day 2: Have students report the results of their surveys. Have the students in their small groups compile the results according to category (e.g., types of music, TV shows, types of food to eat, etc.). Ask students to organize their findings in an appropriate graph. Let each group report the findings to the class.

Distribute the results of the Pew Research Center regarding favorite types of music (see fig. 26.1). Ask students to construct a double bar graph using Microsoft Excel (see fig. 26.2). Give students time to examine and compare the results of their survey with that of the Pew Research Center (see fig. 26.3). After graphing their data (see fig. 26.4), assign students to write their comparisons after discussing them in their small groups.

Musical Genres by Age Percent who say they listen "often" to—		
Music Type	Ages 16–29	Ages 30–49
Rock	45%	42%
Rap or Hip-hop	41%	15%
Rhythm and blues	30%	25%
Country	25%	21%
Classical	14%	12%
Jazz	11%	12%
Salsa or Spanish rock	9%	8%

Fig. 26.1. Musical tastes by age from the Pew Research Center (2009)

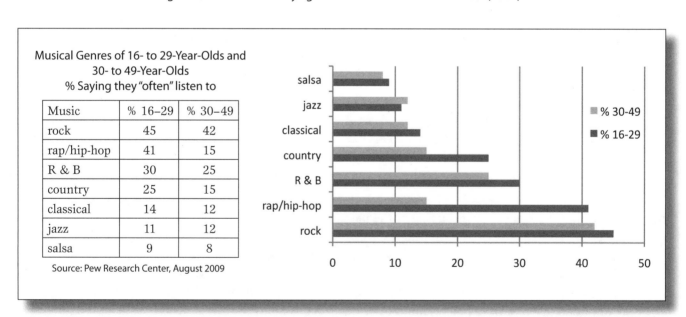

Musical Genres of 16- to 29-Year-Olds and 30- to 49-Year-Olds
% Saying they "often" listen to

Music	% 16–29	% 30–49
rock	45	42
rap/hip-hop	41	15
R & B	30	25
country	25	15
classical	14	12
jazz	11	12
salsa	9	8

Source: Pew Research Center, August 2009

Fig. 26.2. Table and double bar graph of Pew Research Center data

Fig. 26.3. Three seventh graders analyzing data from their survey
to compare with Pew Research Center data

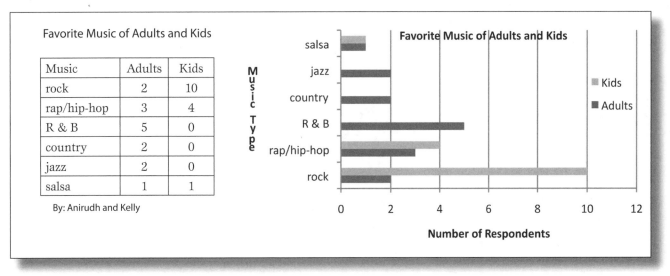

Favorite Music of Adults and Kids

Music	Adults	Kids
rock	2	10
rap/hip-hop	3	4
R & B	5	0
country	2	0
jazz	2	0
salsa	1	1

By: Anirudh and Kelly

Fig. 26.4. Table and double bar graph prepared by seventh graders
using Microsoft Excel

Questions for discussion

1. "How many parents responded to item ---?" [RD]

2. "How many students responded to item ---?" [RD]

3. "Which categories had similar results for parents and students?" [RBW]

4. "Which type of graph was appropriate for the representing the survey data? Why?" [RBY]

5. "How do the results of the class survey compare with the results of the survey conducted by the Pew Research Center?" [RBY]

6. "If you were to conduct the survey again, what would you change? What would you leave the same? Why?" [RBY]

Writing and reading

Using word processing software, have students write a letter to the Pew Research Center describing their survey, the results of their survey, and how their results compared with the results of the

Pew Research Center (see fig. 26.5). Encourage the students to express their opinions and indicate any questions they have about the Pew survey.

```
Dear Pew Research Center:

    I read the results of your music survey and wondered if I
could find out the same thing. My friend Kelly and I wrote a
survey and gave it to 15 adults and 15 kids. We did not ask
ages, but we could tell the difference between adults and kids.
None of the kids like rhythm and blues, jazz, or country music.
Rock and hip-hop are the most popular for kids. You found the
same thing in your survey for the younger people. The adults
in our survey like rhythm and blues the best. The next time we
take a survey, we will ask their ages so we can compare better.

                            Sincerely,
                            Anirudh
```

Fig. 26.5. Letter written by a seventh grader comparing her graph with Pew Research Center data

Activity 27

Topic	Eating lunch at school
Graph forms	Bar graph, line graph
Graph title	From Where Do We Get Our School Lunch?
Objectives	1. To identify ways to obtain school lunch
	2. To collect, organize, and interpret data
	3. To categorize ways of obtaining school lunch
	4. To construct a bar graph
	5. To compare the results of a class, school, or neighborhood survey with the results of a national survey published in a newspaper
	6. To answer comprehension questions on the basis of information in a specific graph
	7. To write a letter to the editor of a newspaper related to constructing and comparing graphs
Vocabulary	Graph, bar graph, line graph, horizontal, vertical, survey, poll, compare, analyze, percent, representative, subsidized, social justice
Materials	Chalkboard and chalk or overhead projector, transparencies, and markers, or document projector; 1-cm graph paper (Appendix 6) or 1/4" graph paper (Appendix 7), pencils, and rulers for each student; survey sheet (similar to Appendix 12); graphs from national surveys (see fig. 27.1 and fig. 27.2)

Procedure

Part 1: Ask students, "How many of you bring your lunch to school?" "How many of you buy your lunch in the cafeteria?" Invite a volunteer to record the responses on the chalkboard, on the overhead projector, or on a document projector. Ask, "What are some other ways of obtaining lunch?" "How many of you obtain your lunch in these other ways?" "How do you think students in other classes would answer these questions? How could we find out?" A survey can be conducted either in the classroom or as an outside assignment. As the students design a plan to conduct the survey, have them discuss any other related questions they might want to ask. Remind them to be sure that they do not collect data from the same respondent more than once.

Once the data have been collected and organized, have students convert the data into equivalent percentages and construct a bar graph. Discuss with the students how to organize the bar graph. Distribute graph paper and have the students draw a set of axes. Do this on the chalkboard, on a transparency, or on a document projector. Depending on the number of pieces of data, it might be necessary to use multiples of two, five, or some other number along the axis labeled "Number of Students (or Respondents)." Review how to write numerals along the axis—that is, stress the meaning of each box and each line. Be sure that the students insert a title on their graphs.

Once the graph is completed, have the students convert the results to percents so that the graph of the survey data can be compared with the graph in figure 27.1. Have students answer questions 1 through 6 below.

Fig. 27.1. Bar graph, *USA TODAY*, 16 September 2009. Reprinted with permission

Part 2: Ask students to analyze the graph in figure 27.2, "How many kids are eating subsidized school lunches?" "How are the graphs in figures 27.1 and 27.2 the same?" "How are the graphs different?" Have students answer questions 7 through 12 below.

Fig. 27.2. Line graph, *USA TODAY,* 1 October 2009. Reprinted with permission

---

Questions for discussion

1. "What is this graph about?" [RD or RBW]

2. "According to the bar graph you made, what percentage of students buy their lunch at school?" [RD]

3. "According to the graph from the newspaper, how does the percentage of students who buy lunch at school nationally compare with the percentage of students who buy lunch locally?" [RBW]

4. "What do you think are other ways that students obtain lunch?" [RBW and RBY]

5. "What questions should we ask about the results reported in the national survey? Do you think it is representative of the entire United States? Why? Why not?" [RBY]

6. "Do you think it is appropriate to compare the class's results with the national survey results? Why? Why not?" [RBY]

7. "What type of a graph is used in figure 27.2?" [RD] "Why do you think this type of graph is used to represent the data in figure 27.2?" [RBY]

8. "What does it mean for lunches to be 'subsidized'?" "Whom do you think subsidizes the lunches?" "Why do you think lunches need to be subsidized?" "Why do you think this is an example of social justice?" [RBY]

9. "How many millions of students were eating subsidized school lunches in 2000?" [RD]

10. "What is the percentage of increase in subsidized school lunches from 1990 to 2008?" [RBY]

11. "In what ways might you be able to use the data in both graphs to make some conjectures about the number or percentage of students who eat subsidized and nonsubsidized school lunches?" "What further information would be needed to confirm your conjecture?" [RBY]

12. "What questions do you have about these graphs?"

Writing and reading

After the students have discussed the answers to the questions, ask them to write a letter to the editor of *USA TODAY*, comparing their graph with the graph in figure 27.1, and pointing out any concerns or questions they have about the published graph. If appropriate, students may also want to refer to the graph in figure 27.2. Ask them to share their work by reading their letters aloud. Encourage them to question their peers.

Activity 28

Topic	Educational opportunities for special-needs children
Graph form	Histogram
Graph title	Enrollment in Programs for the Disabled in the USA, 1994–2006
Objectives	1. To discuss ways schools serve children with special needs
	2. To analyze data
	3. To construct a histogram
	4. To compare a bar graph and a histogram
	5. To answer comprehension questions on the basis of information in a specific graph
	6. To write a letter to the editor of a newspaper related to social justice issues, citing relevant data
Vocabulary	Histogram, trend, social justice
Materials	Chalkboard and chalk, or overhead projector, transparencies, and markers, or document projector; 1-cm graph paper (Appendix 6) or 1/4" graph paper (Appendix 7), pencils and rulers for each student; activity sheet (Appendix 24)

Procedure

Ask students, "How do you think schools serve children with special educational needs?" "Why is it important that schools supply such services for special children?" The number of children from 3 to 21 years of age served annually in educational programs funded by the United States government from 1994 to 2006 is provided in table 28.1 (see Appendix 24).

Table 28.1
Number of Children Ages 3–21 in Special Education Programs Funded by the U.S. Government

Year	Enrollment
1994–95	5,378,000
1995–96	5,572,000
1996–97	5,737,000
1997–98	5,908,000
1998–99	6,056,000
1999–2000	6,195,000
2000–01	6,296,000
2001–02	6,407,000
2002–03	6,523,000
2003–04	6,634,000
2004–05	6,719,000
2005–06	6,713,000

Source: Office of Special Education and Rehabilitative Services, U.S. Department of Education

Discuss features of a histogram and compare with features of a bar graph (see page 5).

Using the data in table 28.1, have students construct a histogram and answer the questions in the following section.

Questions for discussion

1. "What is the graph about?" [RD or RBW]
2. "According to the histogram you made, how many special education students were served in federally funded programs in 2005–2006?" [RD]
3. "Describe the trend that you notice in the histogram." [RBW]
4. "How is the histogram you constructed different from a bar graph?" [RBY]
5. "What is the percent of increase in special education enrollment from 1994–95 to 2004–05?" [RBY]
6. "To what might the increase in enrollment be attributed?" [RBY]
7. "What happened to enrollment from 2004–05 to 2005–06?" [RBW] "What do you think caused this change in enrollment?" [RBY]
8. "Why do you think providing educational opportunities for students with special needs is an example of social justice?" [RBY]
9. "What questions do you have about the data?"

Writing and reading

After the students have discussed the answers to the questions, ask them to write a letter to the editor of any newspaper describing the trends in the data and the continued need for attention to social justice. Ask them to share their work by reading their letters aloud. Encourage them to question their peers.

Activity 29

Topic	"Mystery Graphs" (MSEB 1993; Tierney and Nemirovsky 1992)
Graph form	Picture, bar, circle, and line graphs; double bar and double line graphs
Graph title	Students determine
Objectives	a. Given context-free (i.e., topic-free) picture, bar, circle, line, double bar, and double line graphs—

      1. determine possible situations for which the graphs would be representations;

      2. determine possible titles for the graphs;

      3. label axes appropriately;

      4. write stories about the graphs, describing the data that could have been used to create the graphs.

    b. Given stories and data related to some topics—

      1. sketch graphs that represent the stories and data;

      2. explain how the sketches of the graphs depict the stories and the data.

    c. Given several context-free graphs and possible scenarios depicted by titles—

      1. match the graphs with the titles;

      2. explain how the graphs and the titles are related.

Vocabulary	Picture, bar, circle, line, double bar, and double line graphs; axes; axes labels; increasing, decreasing, key, scale, graph sketch as distinguished from a graph
Materials	a. Appendix 25 ("Here's the Graph—What's the Story?"); chalkboard, document projector, or overhead projector; pencils or pens; lined paper to write stories about each graph
	b. Appendix 26 ("Here's the Story—Where's the Graph?"); enlarged versions of figures 29.1 and 29.2 on chart paper, on the chalkboard, document projector, or on the overhead projector; pencils or pens; straightedge for each student; lined paper to write an explanation about the clues identified to make sketches
	c. Appendix 27 ("Graphs and Titles—Where's the Match?"), pencils or pens; lined paper to write an explanation about the clues identified to make matches

Procedure

    a. Display one of the context-free graphs from Appendix 25 on the chalkboard, document projector, or on the overhead projector. Ask students, "What kind of graph is this?" "What is different about this graph in comparison to the other graphs we have studied and constructed?" [It has no labels, no title, and we can't tell what it's about.] "What could this graph be about?" Graph 1 could represent the number of books three children read in one week where each ideograph of a smiley face represents one book. Graph 2 could represent the height in centimeters of four different children. Graph 3 could represent the portion of the types of books a child read in one month—mystery, biography, science fiction, folktales. Graph 4 could represent distance traveled over time (at a constant speed). Graph 5 could represent the number of books three children read in one week where each ideo-graph of a smiley face represents more than one book. Graph 6 could represent the number of girls (clear bar) and the number of boys (gray bar) who prefer different animals out of three types for pets. Graph 7 could represent the test scores on mathematics and social studies tests for one student during a five-week period of time.

Depending on the students' familiarity with graph types, it may be appropriate to assign a subset of the seven graph tasks in Appendix 25. For example, students who are not familiar with ideographs representing a many-to-one correspondence or multiple bar and line graphs could be assigned Graphs 1–4 only. Students who are more advanced may be assigned Graphs 5–7.

b. Display a graph (e.g., fig. 29.1) and a sketch of the same data (e.g., fig. 29.2) on large chart paper or on the overhead projector. Ask students, "What is the same about these two displays of data?" (e.g., graph type, shape of the data, relationship between or among the data. "What is different about these two displays?" (For example, fig. 29.1 has more detail than fig. 29.2.) Because it does not have all the details of figure 29.1 but preserves the relationships between and among the data, we call figure 29.2 a sketch.

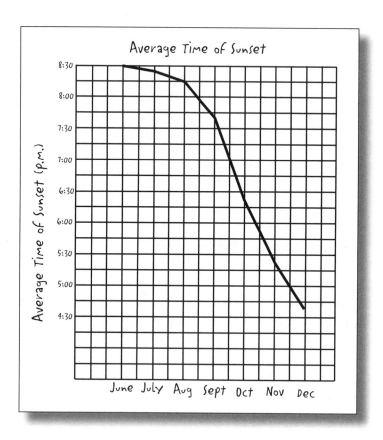

Fig. 29.1. Graph of average time of sunset

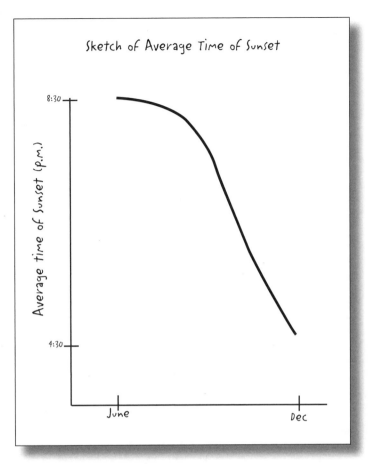

Fig. 29.2. Sketch of average time of sunset

Inform the students, "We are going to be sketching graphs on the basis of information given to us in a story or in a data set." Select one of the stories or data sets from Appendix 26 to work on together with the class. Allow the students to share their thinking about the type of graph that would be appropriate to represent the data and how they would sketch the graph of the data. If appropriate, consider alternative graph forms (e.g., a circle graph is appropriate for Story 1 and Data Set 2; a bar graph or a picture graph is appropriate for Story 2 and Data Set 3; a line graph is appropriate for Data Set 1). After sketching a graph with the class and discussing the characteristics of the data and the features of the graph type, ask students to write an explanation of the clues in the story or the data set that helped them to decide how to make the sketch. Highlight their written work on the chalkboard, document projector, or on the overhead projector. Not all the sketches are meant to be done by each student, but rather, the stories and data sets may be assigned to groups of students to work on and share with the class.

c. Distribute Appendix 27 to the students. Mention to them that they may find more than one title appropriate for several graphs. (Graph 1 could have title (c) or (e); Graph 2 could have title (a), (c) [horizontal axis is the number of students and vertical axis is the number of sit-ups], or (h); Graph 3 could have title (a), (c) [each portion of the circle represents the part of the class able to do a certain number of sit-ups in a minute], or (h); Graph 4 could have title (d). Ask the students to share with the class their reasons for matching the titles with the graphs.

Questions for discussion

Part a: 1.   "What could Graph [insert a graph example] be about?" [RBY]

2.   "What are the clues in Graph [insert a graph example] that helped you determine a situation and a title for the graph?" [RBY]

3.   "What would be an appropriate title for Graph [insert a graph example]?" [RBW]

4.   "What are some questions that can be answered by reading Graph [insert a graph example]?" "What are some questions that cannot be answered by reading Graph [insert a graph example]?" [RBY]

5.   "Without telling the class which graph you pick, tell the story you wrote about one of the graphs. Describe the data that could have been used to create the graph." [RBY] (The class can try to determine which graph the story is about.)

Part b: 6.   "What is a difference between a graph and a sketch of a graph?"

7.   "What are some of the clues that you look for in a story or in a data set to help you determine the type of graph to sketch?" [RWD]

Part c: 8.   "What are the characteristics of the data in each graph that provide clues for matching the graph with appropriate titles?" [RBY]

 # Activity 30

Topic	Balanced nutrition
Graph form	Bar graph
Graph title	Why Don't We Eat More Veggies?
Objectives	1. To discuss reasons for not eating more vegetables
	2. To read a newspaper article with an accompanying graphic display
	3. To collect, organize, and interpret data
	4. To answer comprehension questions on the basis of a newspaper article and a graphic display
	5. To write a letter to the editor
Vocabulary	Bar graph, percent, percent change, billion
Materials	Newspaper article and graphic display (see Appendix 28); chalk and chalkboard, or transparencies, markers, and overhead projector, or document projector; pencils or pens; and lined writing paper for each student; word processing software, computer, printer

Procedure

Ask, "How many of you eat fruits and vegetables as part of your daily diet? What might be some reasons that people do not eat vegetables?" Invite a student volunteer to record the responses on the chalkboard, overhead projector, or document projector.

Distribute the newspaper article, "Not eating your veggies? It's no joke" (see Appendix 28). Ask students to read the article independently and write their answers to the questions for discussion (or a subset of them). Once students are finished reading the article and writing their responses to the questions, have them discuss and compare their responses in small groups of three to four students. Ask one student in each group to record agreed-on responses. Ask another student to record disagreements. Ask a third student in each group to be responsible to report a summary of the group's discussion. At the end of the small-group discussion, each group reporter provides a summary. Ask group reporters to avoid repeating responses of previous groups.

Fig 30.1. *USA TODAY,* 19 October 2009. Reprinted with permission

Questions for discussion

1. "What percentage of the moms surveyed responded that they do not prepare more vegetables for their families because 'family members have different likes and dislikes'?" [RD]
2. "What percent of the respondents indicated that buying and preparing veggies is too time-consuming?" [RD]
3. "What information is given related to the size of the sample of moms polled?" [RD]
4. "Which reason for not eating more veggies occurred least often?" [RBW]
5. "What percentage of the moms responding indicated that fruits and vegetables 'are too expensive'?" [RBW]
6. "Why do you think the sum of the percentages for all of the reasons given is greater than 100 %?" [RBY]
7. "If the bar drawn to represent 46 % is equal to 6.5 cm, how long is the bar that represents 100 %?" [RBY] "Once you find the length of the bar that represents 100 %, determine whether the lengths of the other bars accurately represent the percentages indicated."
8. "What information can be obtained from the text of the newspaper article that is not available in the graph (see fig. 30.1)? What information can be obtained from the graph that is not in the text of the newspaper article? How does the information in the text of the newspaper article compare with the data presented in the graphical display?" [RBY]

Writing and reading

After the students have discussed the answers to the questions, ask them to write a letter to the editor of *USA TODAY,* pointing out any concerns or questions they have about the published graph. Ask them to share their work by reading their letters aloud. Encourage them to question their peers.

References

Arkin, Herbert, and Raymond R. Colton. *Graphs: How to Make and Use Them.* Rev. ed. New York: Harper & Brothers, 1940.

Ash, Russell. *Incredible Comparisons.* Boston: Houghton-Mifflin Co., 1996.

Baer, Edith. *This Is the Way We Go to School: A Book about Children around the World.* New York: Scholastic Books, 1990.

Baratta-Lorton, Mary. *Mathematics Their Way.* Menlo Park, Calif.: Addison-Wesley Publishing Co., 1976.

Becher, Paul G. "Reviewing and Viewing Computer Materials: The Graph Club." *Teaching Children Mathematics* 59 (March 1999): 442.

Belanger, Claude. *I Like the Rain.* Auckland, New Zealand: Shortland Publishing, 1988a.

———. *The T-Shirt Song.* Auckland, New Zealand: Shortland Publishing, 1988b.

Bishop, Ashley, and Sue Bishop. *Teaching Word Analysis Skills.* Huntington Beach, Calif.: Shell Education, 2010.

Bruni, James V., and Helene Silverman. "Graphing as a Communication Skill." *Arithmetic Teacher* 22 (1975): 354–66.

Carpenter, Thomas P., and James M. Moser. "The Acquisition of Addition and Subtraction Concepts in Grades One through Three." *Journal for Research in Mathematics Education* 15 (May 1984): 179–202.

Choate, Laura D., and JoAnn K. Okey. "Graphically Speaking: Primary-Level Graphing Experiences." In *Teaching Statistics and Probability,* 1981 Yearbook of the National Council of Teachers of Mathematics (NCTM), edited by Albert P. Shulte, pp. 33–41. Reston, Va.: NCTM, 1981.

Clines, Francis X. "The City Life: A Most Bookish Borough." *The New York Times,* 17 August 2007, p. A22.

Corwin, Rebecca B., and Susan N. Friel. "Statistics: Sampling." *Elementary Mathematician* 2 (Spring 1988): pullout 1–4.

Curcio, Frances R. "Comprehension of Mathematical Relationships Expressed in Graphs." *Journal for Research in Mathematics Education* 18 (November 1987): 382–93.

———. *Developing Graph Comprehension: Elementary and Middle School Activities.* Reston, Va.: National Council of Teachers of Mathematics, 1989.

———. "Software Review of Graphers." *The Statistics Teacher Network* 42 (Spring 1996): 5–7.

———. "Software Review of Data Explorer." *The Statistics Teacher Network* 49 (Autumn 1998): 1–3.

———. *Developing Data-Graph Comprehension in Grades K–8.* 2nd ed. Reston, Va.: National Council of Teachers of Mathematics, 2001.

Curcio, Frances R., and Alice F. Artzt. "Assessing Students' Ability to Analyze Data: Reaching beyond Computation." *Mathematics Teacher* 89 (November 1996): 668–73.

———. "Assessing Students' Statistical Problem-Solving Behaviors in a Small-Group Setting." In *The Assessment Challenge in Statistics Education,* edited by Iddo Gal and Joan Garfield, pp. 123–137. Amsterdam: International Statistics Institute, 1997.

———. "Students Communicating in Small Groups: Making Sense of Data in Graphical Form." In *Language and Communication in the Mathematics Classroom,* edited by Heinz Steinbring, Maria G. Bartolini Bussi, and Anna Sierpinska, pp. 179–90. Reston, Va.: National Council of Teachers of Mathematics, 1998.

Curcio, Frances R., and Susan Folkson. "Exploring Data: Kindergarten Children Do It Their Way." *Teaching Children Mathematics* 2 (February 1996): 382–85.

Curcio, Frances R., and J. Lewis McNeece. "The Weather Experiment." Unpublished activity. East Elmhurst, N.Y.: Louis Armstrong Middle School, 1987.

"Cutler: It's Up to Us to Teach Students How to Speak Up, Speak Out." *New York Teacher,* 28 November 2009, p. 19.

Davis, Robert B. *Learning Mathematics: The Cognitive Science Approach to Mathematics Education.* Norwood, N.J.: Ablex Publishing Corp., 1984.

diSessa, Andrea A., David Hammer, Bruce Sherin, and Tina Kolpakowski. "Inventing Graphing: Meta-Representational Expertise in Children." *Journal of Mathematical Behavior* 10 (August 1991): 117–60.

Edwards, Lois. Graphers—*Teacher's Guide: Macintosh/Windows*. Pleasantville, N.Y.: Sunburst Communications, 1997.

English, Lyn D., and Donald H. Cudmore. "Using Extranets in Fostering International Communities of Mathematical Inquiry." In *Learning Mathematics for a New Century*, 2000 Yearbook of the National Council of Teachers of Mathematics (NCTM), edited by Maurice Burke, pp. 82–95. Reston, Va.: NCTM, 2000.

Folkson, Susan. "Meaningful Communication among Children: Data Collection." In *Communication in Mathematics, K–12 and Beyond*, 1996 Yearbook of the National Council of Teachers of Mathematics (NCTM), edited by Portia C. Elliott, pp. 29–34. Reston, Va.: NCTM, 1996.

Goldin, Augusta. *Straight Hair, Curly Hair*. New York: HarperCollins, 1966.

Gutstein, Eric, and Bob Peterson, eds. *Rethinking Mathematics: Teaching Social Justice by the Numbers*. Milwaukee, Wis.: Rethinking Schools, 2006.

Huff, Darrell. *How to Lie with Statistics*. New York: W. W. Norton, 1954.

Juraschek, William A., and Nancy S. Angle. "Experiential Statistics and Probability for Elementary Teachers." In *Teaching Statistics and Probability*, 1981 Yearbook of the National Council of Teachers of Mathematics (NCTM), edited by Albert P. Shulte, pp. 8–18. Reston, Va.: NCTM, 1981.

Kamii, Constance, and Ann Dominick. "To Teach or Not to Teach Algorithms." *Journal of Mathematical Behavior* 16, no. 1 (1997): 51–61.

Kirk, Sandra, Paul D. Eggen, and Donald P. Kauchak. "Generalizing from Graphs: Developing a Basic Skill through Improved Teaching Techniques." Paper presented at the annual meeting of the International Reading Association, Saint Louis, Mo., May 1980.

Lappan, Glenda, James T. Fey, William M. Fitzgerald, Susan N. Friel, and Elizabeth Difanis Phillips. *Connected Mathematics: Samples and Populations*. Menlo Park, Calif.: Dale Seymour Publications, 1998.

Landwehr, James, and Ann E. Watkins. *Exploring Data*. Quantitative Literacy Series. Palo Alto, Calif.: Dale Seymour Publications, 1986.

Lee, Jennifer. "Some Schools Can't Afford Hardware and Training." *The New York Times*, 2 September 1999, p. G7.

Lovitt, Charles, and Doug Clarke. *The Mathematics Curriculum and Teaching Program: Professional Development Package. Activity Bank—Volume 1*. Canberra, Australian Capitol Territory, Australia: Curriculum Development Centre, 1988.

Marcovitz, David M. "I Read It on the Computer, It Must be True: Evaluating Information from the Web." *Learning and Leading with Technology* 25 (November 1997): 18–21.

Mathematical Sciences Education Board (MSEB). *Measuring Up*. Washington, D.C.: National Academy Press, 1993.

Mathis, Judi. "Software Reviews—Turning Data into Pictures: Part 1." *Computing Teacher* 16 (October 1988a): 40–48.

———. "Turning Data into Pictures: Part 2." *Computing Teacher* 16 (November 1988b): 7–8, 10.

Markle, Sandra. *Discovering Graph Secrets: Experiments, Puzzles, and Games Exploring Graphs*. New York: Atheneum Books for Young Readers, 1997.

Melser, June, and Joy Cowley. *Poor Old Polly*. Bothwell, Wash.: Thomas C. Wright, 1980.

Moersch, Christopher. "Choose the Right Graph." *Computing Teacher* 22 (February 1995): 31–25.

National Council of Teachers of Mathematics (NCTM). *Curriculum and Evaluation Standards for School Mathematics*. Reston, Va.: NCTM, 1989.

———. *Principles and Standards for School Mathematics*. Reston, Va.: NCTM, 2000.

———. *Curriculum Focal Points: A Quest for Coherence*. Reston, Va.: NCTM, 2006.

Newman, Claire M., and Susan B. Turkel. "The Class Survey: A Problem-Solving Activity." *Arithmetic Teacher* 33 (May 1985): 10–12.

Nuffield Foundation. *Pictorial Representation*. New York: John Wiley & Sons, 1967.

Olivares, Rafael A. *Using the Newspaper to Teach ESL Learners*. Newark, Del.: International Reading Association, 1993.

Pearson, P. David, and Dale D. Johnson. *Teaching Reading Comprehension.* New York: Holt, Rinehart and Winston, 1978.

Pew Research Center. "Forty Years after Woodstock: A Gentler Generation Gap." Released 12 August 2009. Retrieved from pewsocialtrends.org on 22 October 2009.

Reese, Paul, and Mona Monroe. "The Great International Penny Toss." *Learning and Leading with Technology* 24 (March 1997): 28–31.

Royer, Regina. "Teaching on the Internet: Creating a Collaborative Project." *Learning and Leading with Technology* 25 (November 1997): 6–11.

Russell, Susan Jo. "Who Found the Most Shells? (Who Cares?)" *Elementary Mathematician* (1988): 4, 9.

Schwartz, Sydney L. *Teaching Young Children Mathematics.* Westport, Conn.: Praeger, 2005.

Shaw, Jean M. "Let's Do It: Making Graphs." *Arithmetic Teacher* 31 (January 1984): 7–11.

Silverman, Helene. Lesson demonstration at the City College of the City University of New York, July 1987.

———. "Big Ideas from Simple Materials: Playing with Data." *New York State Mathematics Teachers' Journal* 38 (1988): 48–53.

Stone, Antonia. "When is a Graph Worth Ten Thousand Words?" *Hands On!* 11 (Spring 1988): 16–18.

Thomas, Landon, Jr. "Dubai Opens a Tower to Beat All." *The New York Times,* 5 January 2010, pp. B1, B8.

Tierney, Cornelia C., and Ricardo Nemirovsky. "Children's Spontaneous Representations of Changing Situations." *Hands On!* 14 (Fall 1991): 7–10.

———. "Mystery Graphs." *Hands On!* 15 (Spring 1992): 11–14.

Tukey, John. *Exploratory Data Analysis.* Reading, Mass.: Addison-Wesley Publishing Co., 1977.

Van de Walle, John. *Elementary and Middle School Mathematics: Teaching Developmentally.* 4th ed. New York: Addison-Wesley Longman, 2000.

Whitin, David J. "Collecting Data with Young Children." *Young Children* 52 (January 1997): 28–32.

Whitin, David J., Heidi Mills, and Timothy O'Keefe. *Living and Learning Mathematics.* Portsmouth, N.H.: Heinemann, 1990.

Whitin, David J., and Phyllis Whitin. "Learning Is Born of Doubting: Cultivating a Skeptical Stance." *Language Arts* 76 (November 1998): 123–29.

Software

Edwards, Lois. Graphers. Elgin, Ill.: Sunburst Technology, 1996.

———. Data Explorer. Elgin, Ill.: Sunburst Technology, 1998.

Stearns, Peggy Healy. The Graph Club. Watertown, Mass.: Tom Snyder Productions, 1998.

Selected Bibliography

Assad, Saleh. "From Graph to Formula." *Mathematics Teacher* 64 (March 1971): 231–32.

Bright, George W., Wallece Brewer, Kay McClain, and Edward S. Mooney. *Navigating through Data Analysis in Grades 6–8.* Reston, Va.: National Council of Teachers of Mathematics, 2003.

Burns, Marilyn. "Alphabet Math." *Instructor* 97 (September 1987): 48–50.

Chapin, Suzanne, Alice Koziol, Jennifer MacPherson, and Carol Rezba. *Navigating through Data Analysis and Probability in Grades 3–5.* Reston, Va.: National Council of Teachers of Mathematics, 2002.

Chia, David T. "Weather Mathematics: Integrating Science and Mathematics." *Teaching Children Mathematics* 5 (September 1998): 19–22.

Christopher, Leonora. "Graphs Can Jazz Up the Mathematics Curriculum." *Arithmetic Teacher* 30 (September 1992): 28–30.

Collis, Betty. "Learning to Like Social Studies." *Computing Teacher* 15 (April 1988): 30–33.

Curcio, Frances R. "The Effect of Prior Knowledge, Reading and Mathematics Achievement, and Sex on Comprehending Mathematical Relationships Expressed in Graphs." Doctoral diss. New York University, 1981. *Dissertation Abstracts International* 42 (1981): 3047–48A.

————. *The Effect of Prior Knowledge, Reading and Mathematics Achievement, and Sex on Comprehending Mathematical Relationships Expressed in Graphs. Final Report.* Brooklyn, N.Y.: Saint Francis College, 1981. (ERIC Document Reproduction no. ED 210 185).

————. "Incorporating Graphing and Statistics in the Elementary and Middle School Mathematics Curricula." In *Proceedings of Theme Group* 7: Curriculum Towards the Year 2000, edited by John Malone, pp. 83–99. Perth, Western Australia: Curtin University of Technology, 1989.

Curcio, Frances R., and M. Trika Smith-Burke. *Processing Information in Graphical Form.* Paper presented at the annual meeting of the American Educational Research Association, New York, March 1982. Brooklyn, N.Y.: Saint Francis College, 1982. (ERIC Document Reproduction no. ED 215 874).

Dickinson, J. Craig. "Gather, Organize, Display: Mathematics for the Information Society." *Arithmetic Teacher* 34 (December 1986): 12–15.

Dixon, Julie K., and Christy J. Falba. "Graphing in the Information Age: Using Data from the World Wide Web." *Mathematics Teaching in the Middle School* 2 (March–April 1997): 298–304.

Eagle, Edwin. "Toward Better Graphs." *Mathematics Teacher* 35 (March 1942): 127–31.

Friel, Susan N., George W. Bright, and Frances R. Curcio. "Understanding Students' Understanding of Graphs." *Mathematics Teaching in the Middle School* 3 (November–December 1997): 224–27.

Friel, Susan N., Frances R. Curcio, and George W. Bright. "Making Sense of Graphs: Critical Factors Influencing Comprehension and Instructional Implications." *Journal for Research in Mathematics Education* 32 (March 2001): 124–58.

Fry, Edward B. *Graphical Comprehension: How to Read and Make Graphs.* Providence, R.I.: Jamestown Publishers, 1981.

Hannah, Larry. "The Data Base: Getting to Know You." *Computing Teacher* 15 (August–September 1987): 17–18, 41.

Horak, Virginia M., and Willis J. Horak. "Let's Do It: Collecting and Displaying the Data around Us." *Arithmetic Teacher* 30 (September 1982): 16–20.

Kelly, Margaret. "Elementary School Activity: Graphing the Stock Market." *Arithmetic Teacher* 33 (March 1986): 17–20.

Landwehr, James M., Jim Swift, and Ann E. Watkins. *Exploring Surveys and Information from Samples.* Quantitative Literacy Series. Palo Alto, Calif.: Dale Seymour Publications, 1987.

MacDonald-Ross, Michael. "How Numbers Are Shown." *AV Communication Review* 25 (Winter 1977): 359–409.

Nibbelink, William. "Graphing for Any Grade." *Arithmetic Teacher* 30 (November 1982): 28–31.

Russell, Susan Jo, and Susan N. Friel. "Collecting and Analyzing Real Data in the Elementary School Classroom." In *New Directions for Elementary School Mathematics*, 1989 Yearbook of the National Council of Teachers of Mathematics (NCTM), edited by Paul R. Trafton, pp. 134–48. Reston, Va.: NCTM, 1989.

Shaw, Jean M. "Let's Do It: Making Graphs." *Arithmetic Teacher* 31 (January 1984): 7–11.

Sheffield, Linda Jensen, Mary Cavanagh, Linda Dacey, Carol R. Findell, Carole E. Greenes, and Marian Small. *Navigating through Data Analysis and Probability in Prekindergarten–Grade 2.* Reston, Va.: National Council of Teachers of Mathematics, 2003.

Slaughter, Judith P. "The Graph Examined." *Arithmetic Teacher* 30 (March 1983): 41–45.

Sullivan, Delia, and Mary Ann O'Neil. "This Is Us! Great Graphs for Kids." *Arithmetic Teacher* 28 (September 1980): 14–18.

Tufte, Edward R. *The Visual Display of Quantitative Information.* Cheshire, Conn.: Graphics Press, 1983.

Wall, Jennifer J., and Christine C. Benson. "So Many Graphs, So Little Time." *Mathematics Teaching in the Middle School* 15 (September 2009): 82–89.

Whitin, David J. "Dealing with Data in Democratic Classrooms." *Social Studies and the Young Learner* 6 (September–October 1993): 7–9, 30.

Woodward, Ernest, and Frances Byrd. "Make Up a Story to Explain the Graph." *Mathematics Teacher* 77 (January 1984): 32–34.

Appendixes

These materials are also available for download at www.nctm.org/more4u.

1. Supplemental Graph Reading Activities and Answer Key

2. A List of Graph Topics Appropriate for Different Grade Levels

3. How to Make Reusable Teaching Aids

4. Picture Labels for Object Graph and Picture Graph

5. Large-Box Graph Paper

6. 1-cm Graph Paper

7. 1/4" Graph Paper

8. 5-mm Graph Paper

9. Hundredths Disk

10. Daily Activities Data Collection Sheet

11. Picture Graph Activity Sheet

12. Data Collection Sheet

13. Height Data Collection Sheet

14. Fractional Parts of a Circle

15. Comparing Graphic Displays—1

16. Comparing Graphic Displays—2

17. Raisin Experiment Activity Sheet

18. Height over Time Data Collection Sheet

19. Temperature Data Collection Sheet

20. Sunrise and Sunset Data Collection Sheet

21. A Graph Completion Task

22. Average Years of Education of Poorest and Richest 17-to-22-Year-Olds in Selected Countries

23. Comparing Two Different Types of Surveys

24. Analysis of Enrollment of Disabled Students

25. Here's the Graph—What's the Story?

26. Here's the Story—Where's the Graph?

27. Graphs and Titles—Where's the Match?

28. "Not Eating Your Veggies? It's No Joke"

Appendix 1

Supplemental Graph Reading Activities and Answer Key

All questions for Graphs 1–7 are arranged according to the following levels of comprehension: 1–2, Reading the Data; 3–4, Reading between the Data; 5–6, Reading beyond the Data. Answers are given on page 121.

Graph 1: How John Spends His Daily School Allowance

Use the graph above to answer the following questions. Circle the letter of each correct answer.

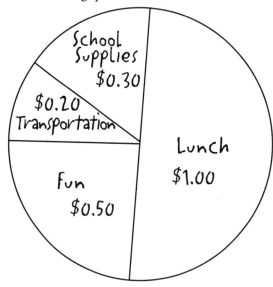

1. What does this graph tell you?
    a. The way John spends his money on weekends
    b. The way John spends his money during one complete week
    c. The way John spends his money for vacations
    d. The way John spends his money for one school day

2. How much does John spend on school supplies for one day?
    a. $0.20
    b. $0.30
    c. $0.50
    d. $1.00

3. What is the total of John's daily school allowance?
    a. $1.90
    b. $2.00
    c. $3.00
    d. $4.00

4. How much more does John spend on lunch than on transportation?
    a. $0.50
    b. $0.70
    c. $0.80
    d. $1.20

5. How much money does John need to pay for lunch for five school days?
    a. $1.00
    b. $2.00
    c. $4.00
    d. $5.00

6. What is the total amount of money John needs for five school days?
    a. $2.00
    b. $7.00
    c. $10.00
    d. $14.00

Graph 2: How Terry Spends a School Day

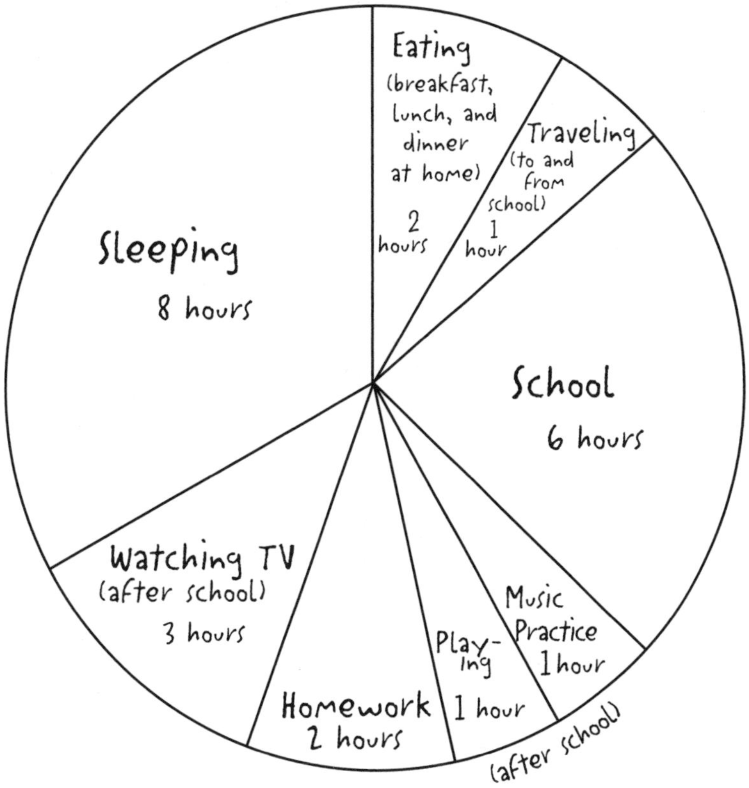

Eating
(breakfast,
lunch, and
dinner
at home)
2 hours

Traveling
(to and
from
school)
1 hour

Sleeping
8 hours

School
6 hours

Watching TV
(after school)
3 hours

Music
Practice
1 hour

Play-
ing
(after school)

Homework 1 hour
2 hours

Use the graph above to answer the following questions. Circle the letter of each correct answer.

1. How many hours in one day does Terry spend in school?
   a. 2 hours
   b. 8 hours
   c. 6 hours
   d. 14 hours

2. For which of the following does Terry spend three hours a day?
   a. Playing after school
   b. Eating
   c. Traveling to and from school
   d. Watching TV

3. What does this graph tell you?
   a. Terry spends the greatest amount of time sleeping.
   b. Terry spends the least amount of time in school.
   c. Terry spends more time playing than watching TV.
   d. Terry spends less time sleeping than in school.

4. What fractional part of a day does Terry spend in school?
   a. 1/12
   b. 1/4
   c. 1/3
   d. 1/2

5. How many hours a week (not including Saturday and Sunday) does Terry spend on homework?
   a. 2 hours
   b. 8 hours
   c. 10 hours
   d. 14 hours

6. What fractional part of a week (not including Saturday and Sunday) does Terry spend sleeping?
   a. 1/2
   b. 1/12
   c. 1/5
   d. 1/3

Graph 3: Height of the Rodriguez Children in March

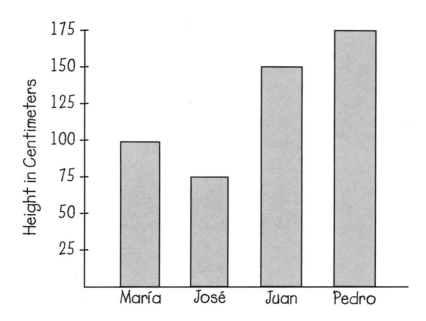

Use the graph above to answer the following questions. Circle the letter of each correct answer.

1. What does this graph tell you?
    a. The weights of the four Rodriguez children in March
    b. The grades of the four Rodriguez children in March
    c. The heights of the four Rodriguez children in March
    d. The ages of the four Rodriguez children in March

2. How tall was María?
    a. 75 inches
    b. 100 inches
    c. 100 centimeters
    d. 125 centimeters

3. Who was the tallest?
    a. Juan
    b. Pedro
    c. José
    d. María

4. How much taller was Juan than José?
    a. 25 centimeters
    b. 50 centimeters
    c. 75 inches
    d. 75 centimeters

5. If María grows 5 centimeters and José grows 10 centimeters by the following September, who will be taller, and by how much?
    a. María will be taller by 20 centimeters.
    b. José will be taller by 20 centimeters.
    c. María will be taller by 5 centimeters.
    d. José will be taller by 5 centimeters.

6. If Pedro is 5 years old, which of the following is a correct statement?
    a. Pedro is much too short for his age.
    b. Pedro could never be that tall for his age.
    c. Pedro is of average height for his age.
    d. Pedro is thin for his age.

Graph 4: The Number of Children in Mr. Kahn's Class Celebrating a Birthday during Each Month of the Year

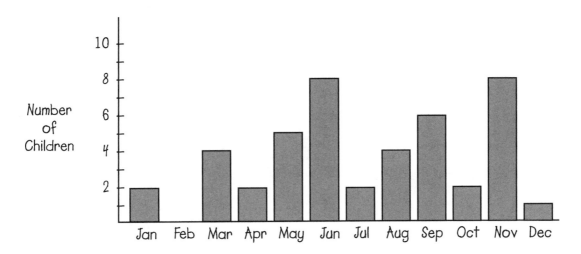

Use the graph above to answer the following questions. Circle the letter of each correct answer.

1. How many children celebrate a birthday in February?
   - a. 0
   - b. 1
   - c. 2
   - d. 4

2. During which month are there eight children who celebrate their birthdays?
   - a. May
   - b. July
   - c. September
   - d. November

3. What does this graph tell you?
   - a. There are more birthdays during June and November than during any other month of the year.
   - b. There are more birthdays during May than during any other month of the year.
   - c. There are fewer birthdays during June and November than during any other month of the year.
   - d. As the year progresses from January to December, the number of birthdays decreases.

4. How many children are in Mr. Kahn's class?
   - a. 10
   - b. 30
   - c. 44
   - d. 55

5. What is the probability that the birthday being celebrated in December occurs on 25 December?
   - a. 1/31
   - b. 25/31
   - c. 30/31
   - d. 1

6. Sally's birthday is in February. According to the graph, which of the following statements is correct?
   - a. Sally was probably born on 29 February.
   - b. Sally is not in Mr. Kahn's class.
   - c. Sally is in Mr. Kahn's class.
   - d. Sally is the only one celebrating a birthday in February.

Graph 5: Average Time of Sunset

Use the graph to answer the following questions. Circle the letter of each correct answer.

1. What is the average time that the sun sets in October?
   - a. 5:00 p.m.
   - b. 5:15 p.m.
   - c. 6:15 p.m.
   - d. 7:30 p.m.

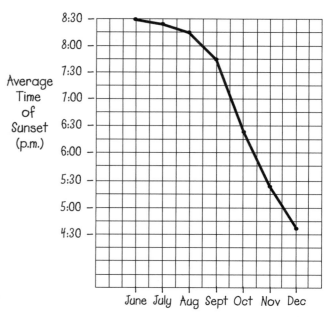

2. The average time of sunset is 4:35 during which month?
   - a. October
   - b. November
   - c. December
   - d. January

3. As the months progress from June to December, which of the following is true about the average time of sunset?
   - a. It gets earlier
   - b. It gets later
   - c. It remains the same
   - d. It first gets earlier and then later

4. How much longer do you have to play outside (before it gets dark) in July than you have in October?
   - a. 1 1/4 hours
   - b. 1 1/2 hours
   - c. 2 hours
   - d. 3 hours

6. As the months progress from June to December, the average time of sunrise gets later. What do you expect to happen to the average number of daylight hours during this time?
   - a. Increases
   - b. Decreases
   - c. Remains the same
   - d. First decreases, and then increases

5. Which of the following graphs represents the average time of sunset from January to June that would make the graph above represent one complete year?

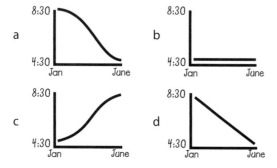

Graph 6: The Number of Books the Jones Children Read per Month

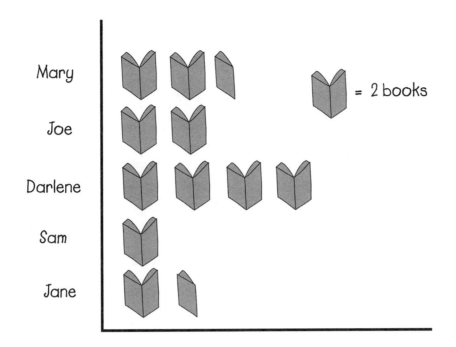

Use the graph above to answer the following questions. Circle the letter of each correct answer.

1. How many books did Sam read in one month?
    a. 0
    b. 1
    c. 1 1/2
    d. 2

2. Who read 2 1/2 books in one month?
    a. Mary
    b. Jane
    c. Joe
    d. No one

3. Who read the fewest number of books in one month?
    a. Darlene
    b. Sam
    c. Jane
    d. Joe

4. In one month, how many more books did Darlene read than Jane?
    a. 1 1/2
    b. 2 1/2
    c. 5
    d. 5 1/2

5. At the end of one year, about how many books will Joe have read?
    a. 16
    b. 24
    c. 30 1/2
    d. 48

6. About how many books does Darlene read in one week?
    a. 1
    b. 2
    c. 4
    d. 8

Graph 7: Stamps Collected by Children

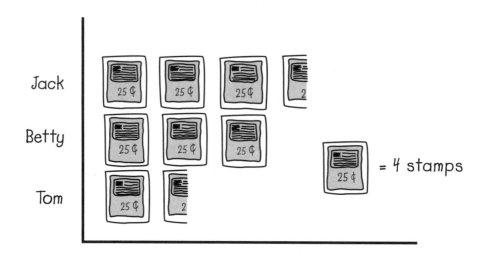

Use the graph above to answer the following questions. Circle the letter of each correct answer.

1. What does each symbol represent?
   a. One stamp
   b. One-half stamp
   c. Four stamps
   d. Twenty-five stamps

2. How many stamps has Tom collected?
   a. 1 1/2
   b. 6
   c. 37 1/2
   d. 50

3. How many more stamps has Betty collected than Tom?
   a. 1 1/2
   b. 3
   c. 6
   d. 37 1/2

4. According to the graph, which of the following statements is true?
   a. Jack has 1/2 more stamps than Betty
   b. Tom has 8 fewer stamps than Jack
   c. Betty has 1 1/2 more stamps than Tom
   d. Betty has 75¢ worth of stamps

5. If the children sold their stamps for the value on the face of the stamp, who, if any of the children, would be able to buy a new bicycle using this money?
   a. Tom
   b. Betty
   c. Jack
   d. None of the children

6. One of Tom's stamps is rare, but he doesn't know it. He and Betty are going to trade one stamp for one stamp. What is the probability that Betty receives Tom's rare stamp?
   a. 1/2
   b. 1/12
   c. 1/6
   d. 1/18

Supplemental Graph Reading Activities Answer Key

Graph 1	Graph 2	Graph 3	Graph 4
1. d	1. c	1. c	1. a
2. b	2. d	2. c	2. d
3. b	3. a	3. b	3. a
4. c	4. b	4. d	4. c
5. d	5. c	5. a	5. a
6. c	6. d	6. b	6. b

Graph 5	Graph 6	Graph 7
1. c	1. d	1. c
2. c	2. d	2. b
3. a	3. b	3. c
4. c	4. c	4. b
5. c	5. d	5. d
6. b	6. b	6. c

Appendix 2

A List of Graph Topics Appropriate for Different Grade Levels

The graph topics were collected from various sources (see Nuffield 1967; Shaw 1984). The topics were then rated by twenty-four elementary and middle school teachers, with each having over five years of teaching experience at multiple grade levels. Grade levels were agreed on by at least 75 percent of the teachers.

Topic and Question	Grade Level								
	K	1	2	3	4	5	6	7	8
Favorites: What/who is your favorite...									
color?	x	x	x	x					
dinosaur?	x	x	x	x					
fairy-tale character?	x	x	x	x					
toy?	x	x	x	x	x				
type of pet?	x	x	x	x	x				
zoo animal?	x	x	x	x					
holiday?	x	x	x	x	x	x	x	x	x
ice cream flavor?	x	x	x	x	x	x	x	x	x
candy?	x	x	x	x	x	x	x	x	x
snack?	x	x	x	x	x	x	x	x	x
dessert?	x	x	x	x	x	x	x	x	x
day of the week?		x	x	x					
chore at home?			x	x					
song?			x	x	x	x	x	x	x
(nonboard) game?			x	x	x	x	x	x	x
type of food?			x	x	x	x	x	x	x
vegetable?			x	x	x	x	x		
fruit?			x	x	x	x	x		
TV show?			x	x	x	x	x	x	x
classroom job?			x	x	x	x	x		
season of the year?			x	x	x	x	x	x	x
subject in school?				x	x	x	x	x	x
fast-food restaurant?				x	x	x	x	x	x
sport to play?					x	x	x	x	x
board game?					x	x	x	x	x
book (title)?					x	x	x	x	x
story (title)?					x	x	x	x	x
(least favorite) subject (in school)?					x	x	x	x	x
pizza topping?					x	x	x	x	x
school activity?					x	x	x	x	x
flower?					x	x	x	x	x
videogame?					x	x	x	x	x
soft drink?					x	x	x	x	x

Topic and Question	Grade Level								
	K	1	2	3	4	5	6	7	8
Favorites (continued): What/who is your favorite ...									
weekend activity?						x	x	x	x
sport to watch?						x	x	x	x
book type?						x	x	x	x
movie (title)?						x	x	x	x
hobby?						x	x	x	x
lucky number?						x	x	x	x
type of apple (e.g., raw, sauce, juice)?						x	x	x	x
insect?						x	x	x	x
rock group?							x	x	x
male singer?							x	x	x
female singer?							x	x	x
male athlete?							x	x	x
female athlete?							x	x	x
athletic shoe brand?							x	x	x
musical instrument?							x	x	x
TV soap opera?							x	x	x
car?							x	x	x
computer?							x	x	x
brand of toothpaste?							x	x	x
author?							x	x	x
radio station?							x	x	x
designer label?								x	x
actor?								x	x
actress?								x	x

Topic and Question	Grade Level								
	K	1	2	3	4	5	6	7	8
Counting, Quantity: How many ...									
teeth have you lost?	x	x	x						
children drink milk for lunch?	x	x	x	x					
children are in your family?	x	x	x	x	x	x			
children in our class have the same name?		x	x	x					
children did not finish their milk at lunch?		x	x	x					
TVs do you have in your house?		x	x	x	x	x			
letters in your last name?			x	x					
children present or absent today?			x	x					
vowels (or consonants) in your first (or last) name?			x	x					
coins of each type do we have?			x	x					
children wear (or don't wear) glasses?		x	x	x	x				

Topic and Question	Grade Level								
	K	1	2	3	4	5	6	7	8
Counting, Quantity (continued): How many …									
sunny days did we have in each week (for one month)?			x	x	x	x	x		
children celebrate a birthday during each month?			x	x	x	x	x		
children celebrate a birthday during each season?				x	x	x	x		
children are right- (or left-) handed?				x	x	x			
windows do you have in your house?				x	x	x			
doors does your house have?				x	x	x			
mirrors do you have in your house?				x	x	x			
rooms does your house have?				x	x				
books have you read in one week (or month)?					x	x	x	x	x
times are the letters of the alphabet used in a 100-word passage?						x	x	x	x
parts of speech (noun, verb, adjective, adverb, other) are in a "random" sentence?						x	x	x	x
riders are there in cars passing the school between 10:00 a.m. and 11:00 a.m.?						x	x	x	x
children did each U.S. president have?						x	x	x	x
times does a vowel appear in one newspaper article?						x	x	x	x
vowels occur in ten lines of prose?						x	x	x	x
kilowatt-hours of energy (i.e., electricity) do we use each month?							x	x	x
calories in (some) milk products?							x	x	x
calories in seven favorite foods?							x	x	x

Topic and Question	Grade Level								
	K	1	2	3	4	5	6	7	8
Categorizing, Measurement: What is/are …									
your bedtime?		x	x	x					
your hair color?		x	x	x	x	x			
your eye color?		x	x	x	x	x			
the types of food we eat for lunch?			x	x	x	x	x	x	
your shoe size?				x	x	x	x	x	
the indoor/outdoor temperature at noon for five days?				x	x	x	x	x	x
type of footwear you have on?				x	x	x	x	x	x
the national average height/weight of children ages seven, eight, and nine?					x	x	x	x	x

Topic and Question	Grade Level								
	K	1	2	3	4	5	6	7	8
Categorizing, Measurement (continued): What is/are …									
the kinds of cars your parents drive/own?					x	x	x	x	x
the size of books we have on our desks?					x	x	x		
the distance you can throw a softball, basketball, etc.?					x	x	x	x	x
the length of our hands compared to the length of our feet?						x	x	x	x
the height of the tallest child in each of the _____ (th) grades?						x	x	x	x
maximum/minimum temperatures for five consecutive days?						x	x	x	x
the population of (selected) countries?						x	x	x	x
the population of our neighboring communities?						x	x	x	x
the length of the five longest rivers in the world?						x	x	x	x
the life span of (selected) animals?						x	x	x	x
our community/city population (over a given period of time?						x	x	x	x
the number of sit-ups you can do in one minute?						x	x	x	x
the distance you can run in one minute?						x	x	x	x
the approximate area of your (dominant) hand?							x	x	x
the length of our feet compared to our heights?							x	x	x
the stopping distance (in feet) for various car speeds?							x	x	x
the cost of (selected) kinds of cars?							x	x	x
the approximate mile/gallon for (selected) kinds of cars?							x	x	x
the average time of sunrise/sunset for each month of the year?							x	x	x
the approximate land areas of (selected) countries?							x	x	x
the population of (selected) cities?							x	x	x
the height of (selected) mountains?							x	x	x
the barometric pressure recorded at different times during the day?							x	x	x
the depths of the oceans of the world?							x	x	x
the length of the sun's shadow at different times during the day (during different seasons)?							x	x	x

Topics and Questions	Grade Level								
	K	1	2	3	4	5	6	7	8
Categorizing, Measurement (continued): What is/are ...									
the number of centimeters a candle will burn in a given amount of time?							x	x	x
the frequency of the letters used in the English language?								x	x
American casualties resulting from (selected) wars?								x	x

Topics and Questions	Grade Level								
	K	1	2	3	4	5	6	7	8
Miscellaneous, Measurement, Categorizing ...									
How do we travel to school?	x	x	x	x	x	x			
What type of pet do you own?	x	x	x	x	x	x	x		
How tall are you? (over a period of time)?		x	x	x	x	x	x	x	
How tall is each child in our class?				x	x	x	x		
How much does each child weigh?				x	x	x	x		
In what type of dwelling do you live?				x	x	x	x		
How do you spend each hour of a school day?					x	x	x	x	x
How do you spend each hour on a Saturday?					x	x	x	x	x
How do you spend your daily school allowance?					x	x	x	x	x
How much milk does each class consume in one day?					x	x	x	x	
How does your height/weight change over a period of four months?						x	x	x	x
How fast does your plant grow (result of seed planting)?						x	x	x	x
What is the relationship between height and weight?							x	x	x
How does the distance we can jump (broad/long jump) compare with our height?							x	x	x
How much water do we consume in one month?							x	x	x
What were the European countries represented by explorers who traveled to the New World between 1492 and 1693?							x	x	x
What was our city's public school enrollment from 1900 to 1980?							x	x	x
How do (selected) stock prices change over a period of time?								x	x

Appendix 3

How to Make Reusable Teaching Aids

Floor Grid (for people graphs in Activities 3 and 4—see fig. A3.1)

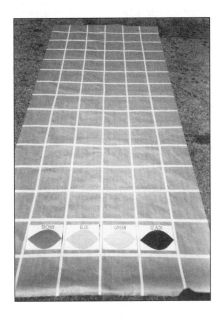

Fig. A3.1

Materials

Heavy brown packing paper, oilcloth, or heavy plastic; yardstick; 1" or 3/4" masking tape or cloth tape; pencil, marker, or paint

Directions

a.   Measure a 6'6" × 8'8" piece of paper, oilcloth, or plastic. If the paper is only 3' wide, you will have to attach two 3' widths together or just make a narrower floor grid. If you use a smaller width, the choices will be limited to approximately three categories.

b.   Mark off 1' × 1' squares using 1" masking tape, cloth tape, or markers to make the horizontal and vertical lines.

c.   Depending on the quality and durability of the material used, you may have to cover the floor grid with clear contact paper.

From Choate and Okey (1981, pp. 34–37)

Object Grid (for object and picture graphs in Activity 6—see fig. A3.2)

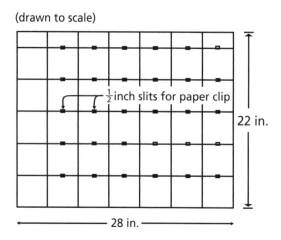

Fig. A3.2

Materials

22" × 28" piece of white oak tag or tagboard, clear contact paper, ruler, pencil, permanent black marker

Directions

a. Measure 4" × 4" squares with a pencil.

b. Draw over the horizontal and vertical lines with a permanent black marker.

c. Cover the grid with clear contact paper to protect the surface.

d. To use the grid for a picture graph, cut a small 1/2" slit in the middle of the top edge of each square. A paper clip should be inserted to hold small pictures or drawings for the picture graph.

Appendix 4

Picture Labels for Object Graph and Picture Graphs

Appendix 5

Large-Box Graph Paper

Appendix 6

1-cm Graph Paper

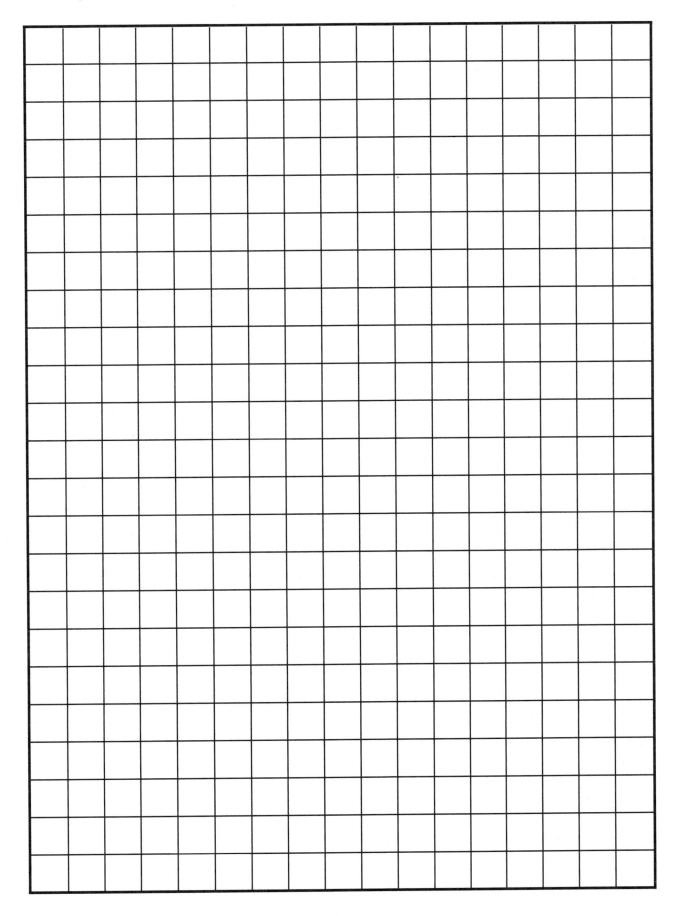

Appendix 7

1/4" Graph Paper

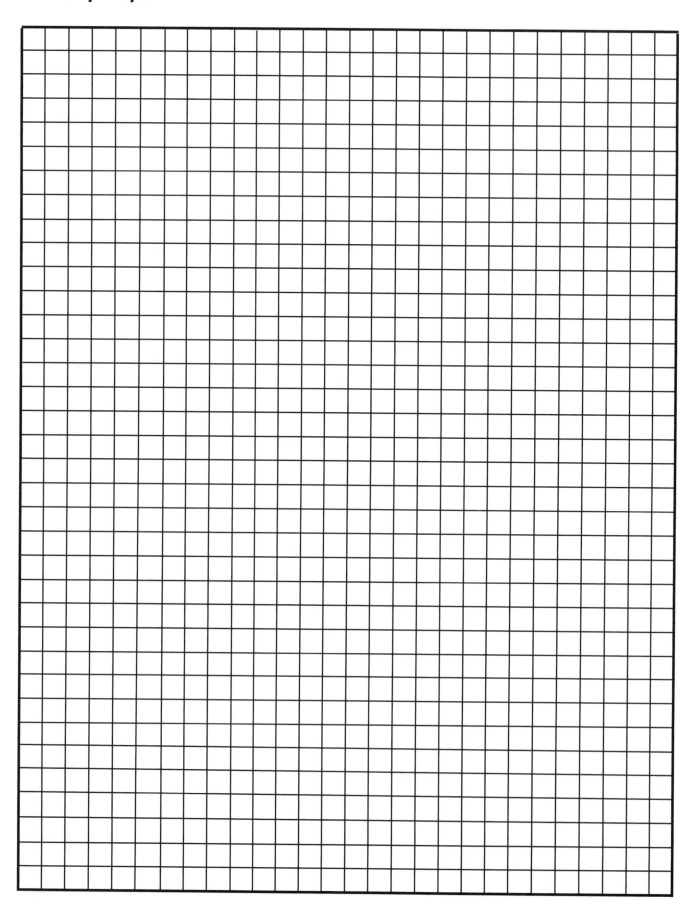

Appendix 8

5-mm Graph Paper

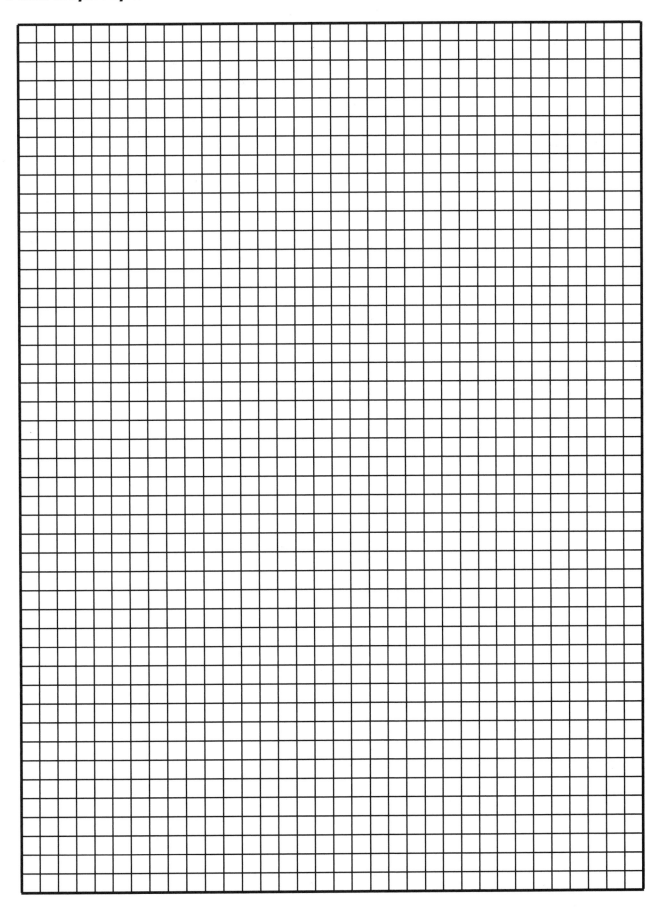

Appendix 9

Hundredths Disk

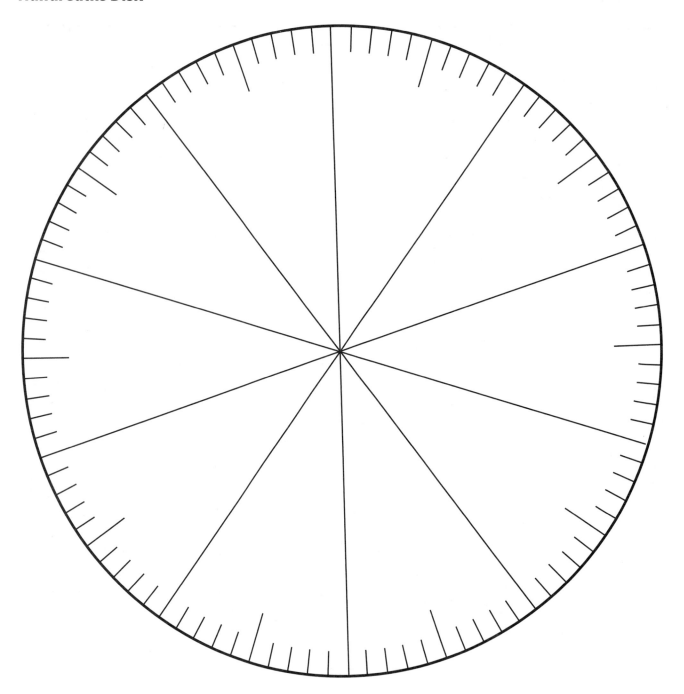

Appendix 10

Daily Activities Data Collection Sheet

Name _____ Date _____

School _____ Class _____

1. Fill in each hour time slot with an activity. Think of a typical Saturday when filling in the activity.

A.M		P.M.
12:00 _____		12:00 _____
1:00 _____		1:00 _____
2:00 _____		2:00 _____
3:00 _____		3:00 _____
4:00 _____		4:00 _____
5:00 _____		5:00 _____
6:00 _____		6:00 _____
7:00 _____		7:00 _____
8:00 _____		8:00 _____
9:00 _____		9:00 _____
10:00 _____		10:00 _____
11:00 _____		11:00 _____

2. List each activity and the total number of hours spent engaged in the activity.

Activity	Number of Hours
_____	_____
_____	_____
_____	_____
_____	_____
_____	_____
_____	_____
_____	_____
_____	_____
_____	_____
_____	_____
_____	_____

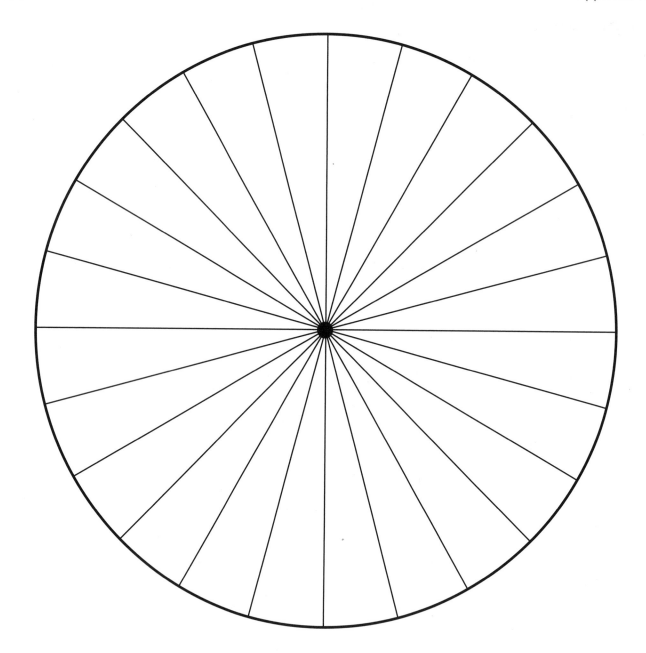

3. Using each space to represent one hour, label each activity in the graph. Use a different color to highlight each activity.

   a. How many hours do you spend sleeping? Eating? Watching TV?

   b. For which activity do you spend the most amount of time?

   c. For which activity do you spend the least amount of time?

   d. After analyzing how you spend your time on Saturday, is there anything you would like to change? If so, what and how?

   e. How would a weekday graph be different from this one?

   f. How would a summer day graph be the same or different as compared to this graph?

   g. Think of a question that you can ask.

Appendix 11

Picture Graph Activity Sheet

Name ————————————————— Date —————————————————

School ————————————————— Class —————————————————

Graph 1. Picture Graph (1 symbol represents 1 student)

Graph 2. Picture Graph (1 symbol represents 2 students)

Graph Title: —————————————

Graph Title: —————————————

Number of children in family

Number of children in family

Legend:

1. How many children in our class have two children in their families?

2. How many children are in the most number of families?

3. How many children are in the fewest number of families?

Legend:

4. How are the two picture graphs the same?

5. How are the two picture graphs different?

Appendix 12

Data Collection Sheet

Name _____ Date _____

School _____ Class _____

1. Record the heights of three adults and the lengths of their bare right feet, to the nearest centimeter.

             #1   #2   #3

     Height _____   _____   _____

Foot Length _____   _____   _____

2. Record the favorite ice cream flavors of five persons, including yourself. There are two ways; try both. First, simply ask for their favorite flavor. Second, ask for their favorite among chocolate, strawberry, vanilla, and other.

   _____   _____   _____   _____   _____
     #1   #2   #3   #4   #5

3. Record the eye colors of five persons, including yourself.

      _____   _____   _____   _____
        blue   brown   gray   other

4. Record the hair colors of five persons, including yourself.

   _____   _____   _____   _____   _____   _____
     blond   brown   black   red   gray   other

5. Record the numbers of letters and spaces in the names of five persons, including yourself.

      _____   _____   _____   _____   _____

6. Record the numbers of each type of coin in your possession right now. (List denomination under each.)

      _____   _____   _____   _____   _____

7. Record the favorite TV programs of five persons, including yourself.

   _____   _____

   _____   _____

   _____   _____

8. Record the birth months of five persons, including yourself.

      _____   _____   _____   _____   _____

Adapted from Juraschek and Angle (1981).

Appendix 13

Height Data Collection Sheet

Name ———————————————   Date ————————————————

School ——————————————   Class ————————————————

1. Record the heights of you and your friends.

Name	Height (in inches or centimeters)

Construct a bar graph using the data above. Write a story about your graph.

2. Record the heights and long-jump distances for you and your friends.

Name	Height (in inches)	Long Jump (in inches)

Construct a double bar graph using the data recorded above. Write a story about your graph.

Appendix 14

Fractional Parts of a Circle

Name _____ Date _____

School _____ Class _____

Fraction	Division representation of fraction	Decimal representation of fraction	Diagram	Number of degrees in fractional part of circle
$\frac{1}{2}$				
$\frac{1}{4}$				
$\frac{1}{3}$				
$\frac{2}{3}$				
$\frac{1}{6}$				
$\frac{5}{6}$				

Appendix 15

Comparing Graphic Displays—1

Name _____ Date _____

School _____ Class _____

DIRECTIONS: Read the following three displays of information.

1. Write as many things as you can about the information given.

2. Which of these displays did you find easiest to understand? Why? Would you like all information to be presented in this way?

3. If you think any of these displays can be changed so that they are easier to understand, describe how you would change them.

The Number of Children in Mr. Kahn's Class
Celebrating a Birthday during Each Month of the Ye

Month	Number of Children
January	2
February	0
March	4
April	2
May	5
June	8
July	2
August	4
September	6
October	2
November	8
December	1

DISPLAY 1

The Number of Children in Mr. Kahn's Class
Celebrating a Birthday during Each Month of the Year

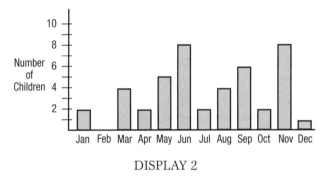

DISPLAY 2

The Number of Children in Mr. Kahn's Class
Celebrating a Birthday during Each Month of the Year

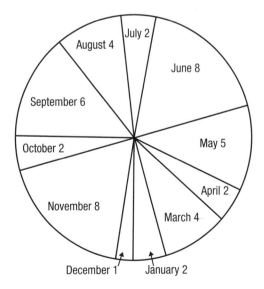

DISPLAY 3

Appendix 16

Comparing Graphic Displays—2

DIRECTIONS: Read the following three displays of information.

1. Write as much as you can about the information given in each display.

2. Which of these displays did you find easiest to understand? Why? Would you like to have all information presented in this way?

3. If you think any of these displays can be changed so that they are easier to understand, describe how you would change them.

Display 1

Types of Homes	Number of People
Single-family house	140
Apartment	32
Two-family house	12
Co-op or condominium	6
Trailer home	6
Town house	4

Display 2

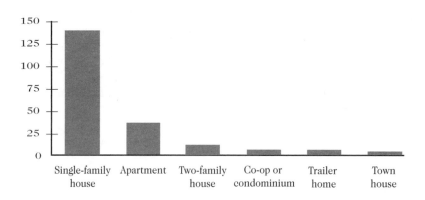

Display 3

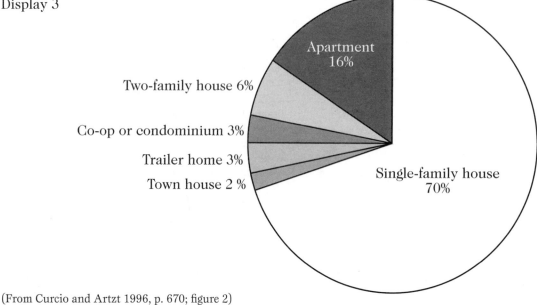

(From Curcio and Artzt 1996, p. 670; figure 2)

Appendix 17

Raisin Experiment Activity Sheet

Name _____ Date _____

School _____ Class _____

1. Estimate how many raisins are in your raisin box. Estimate:

2. Record the estimates of all the students in your group in the table below.

3. What is the lowest estimate?

4. What is the highest estimate?

5. Which estimate occurs most often?

6. Count the raisins in your box. How many are there?

7. Record the actual number of raisins in the boxes of all the students in your group. Use the table below.

8. What is the fewest number of raisins?

9. What is the greatest number of raisins?

10. How many raisins occurred most often?

Name of Student	Estimate	Actual Count

11. Construct a double bar graph.

12. Answer the following—

   a. How many raisins did you have in your box? How close was the actual count to your estimate?

   b. Whose estimate was the closest to the actual count?

   c. What brand of raisins did we use?

   d. What is another brand of raisins we could use? How do you think a box of the same size would compare?

   e. Think of a question that could be answered using your graph.

13. Write a story about the activity and your graph.

Appendix 18

Height over Time Data Collection Sheet

Name _____ Date _____
School _____ Class _____

Keep a record of your height (or the height of a plant) for a given period of time.

Date	Height (in cm)

Construct a line graph using the data above. Write a story about your graph.

Appendix 19

Temperature Data Collection Sheet

Name _____ Date _____

School _____ Class _____

Keep a record of the a.m. and p.m. temperature. Be sure to check the temperature at the same times each day during the next seven days.

Temperature (in Fahrenheit)

Date	A.M.	P.M.

1. Construct a line graph using either the a.m. or p.m. temperature.

2. Construct a multiple line graph using the data above. For each graph, write a short story.

Appendix 20

Sunrise and Sunset Data Collection Sheet

Name _____ Date _____

School _____ Class _____

Keep a record of the time of sunrise and sunset for the next fourteen days. You can obtain this information in the daily newspaper.

Day and Date	Time of Sunrise	Time of Sunset
Mean		
Median		

Construct a multiple line graph using the data above. Write a story about your graph.

Appendix 21

A Graph-Completion Task

PART 1. Read the following two displays of information. Working alone, write as much as you can about the information given.

Share the interpretation of the information with the members of your group. The recorder of the group should write a statement that represents the group's interpretation.

Average Time of Sunset

Month	Time
June	8:30 P.M.
July	8:25 P.M.
August	8:15 P.M.
September	7:40 P.M.
October	6:20 P.M.
November	5:20 P.M.
December	4:35 P.M.

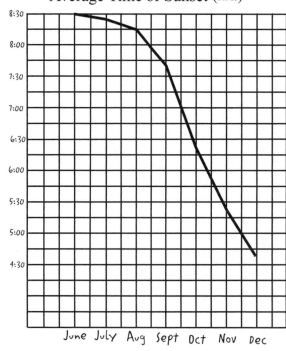

Average Time of Sunset (P.M.)

PART 2. Work with the members in your group. Using the line graph entitled "Average Time of Sunset," draw a picture of what the graph would look like if it were to continue from January to May.

(From Curcio and Artzt 1996, p. 671; Curcio and Artzt 1997, p. 125; Curcio and Artzt 1998, p. 181)

Appendix 22

Name _____   Date _____

School _____   Class _____

Average Years of Education of Poorest and Richest 17-to-22-Year-Olds
in Selected Countries, 2005

1. Locate each country listed in the table on a map of the world. Identify the continent in which the country is located and insert it in the space provided in the table.

Country	Continent	Average Number of Years Poorest[1] Students Are in School	Average Number of Years Richest[1] Students Are in School
Bangladesh		3.7	8.1
Burkina Faso		0.75	5.8
Ghana		3	9
Guatemala		1.9	8.3
India		4.5	11
Mozambique		1.9	5
Nicaragua		2.5	9
Nigeria		3.9	10
Peru		6.5	11
Philippines		6.3	10.9
Tanzania		3.9	8.1

[1] Poorest 20 % of the population, richest 20 % of the population
Source: UNESCO

2. Examine the data in the table. How do the average number of years in school for the poorest and richest students in each country compare?

3. Working in groups of two, enter the data in a computer and construct a double bar graph, comparing the average number of years in school for the poorest and richest students. For the headings of each column use "Country," "Av # of Yrs Poorest Attend School," "Av # of Yrs Richest Attend School."

4. Compare the double bar graph constructed on the computer with the data in the table. How are they the same? How are they different? Which is easier to read? Why?

5. Answer the following questions:
    a. What is this graph about?

    b. Why do you think this graph is called a "double bar graph"? What are the characteristics of a double bar graph?

    c. What is the average number of years that the poorest students in Ghana spend in school?

    d. What is the average number of years that the richest students in Peru spend in school?

    e. In which country is the average number of years spent in school the least for the poorest students?

    f. In which country is the average number of years spent in school the greatest for the richest students?

    g. Of the countries listed, in which continent, on average, do the poorest students spend the least amount of time in school?

    h. Of the countries listed, in which continent, on average, do the richest students spend the greatest amount of time in school?

    i. What do the data tell us about children's opportunity for education and poverty?

    j. What are the advantages of reading data in a double bar graph? What are the advantages of reading data in a table?

    k. Why do you think there are differences between the average number of years that the poorest and richest students spend in school in each of the countries listed in the table?

    l. How do you think public education in the United States attempts to provide equal opportunities among the poorest and richest students?

    m. How can we determine whether public education is equitable among poor and rich students?

    n. Think of a question you can ask about this graph.

    o. Write a letter to UNESCO using the data to recommend equal educational opportunities for all children, poor as well as rich, throughout the world. Share a draft of your letter with your partner and critique each other's work.

Appendix 23

Comparing Two Different Types of Surveys

Analyze each pair of questions and identify at least one advantage and one disadvantage of using each format (open ended vs. controlled).

Open-ended Format	**Controlled Format**
1. What is your favorite school subject?	1. What is your favorite school subject? a. Language Arts b. Science c. Math d. Social Studies e. Music f. Art h. Other: _____
2. What is your favorite sport to play?	2. What is your favorite sport to play? a. Baseball b. Soccer c. Football d. Tennis e. Other: _____
3. What is your favorite type of sport to watch?	3. What is your favorite type of sport to watch? a. Baseball b. Soccer c. Football d. Tennis e. Other: _____
4. What is your favorite type of music?	4. What is your favorite type of music? a. Hip Hop b. Country c. Rock d. Rap e. Classical f. Pop g. Salsa h. Other: _____

Appendix 24

Analysis of Enrollment of Disabled Students

Name _____ Date _____

Analysis of Enrollment of Disabled Students in the USA, 1994-2006

1. Using the data in the table below, construct a histogram.

Number of Children Ages 3-21 in Special Education Programs
Funded by the U.S. Government

Year	Enrollment
1994-95	5,378,000
1995-96	5,572,000
1996-97	5,737,000
1997-98	5,908,000
1998-99	6,056,000
1999-2000	6,195,000
2000-01	6,296,000
2001-02	6,407,000
2002-03	6,523,000
2003-04	6,634,000
2004-05	6,719,000
2005-06	6,713,000

Source: Office of Special Education and Rehabilitative Services,
U.S. Department of Education

2. Answer the following questions.
    a. What is the graph about?

    b. According to the histogram you made, how many special education students were served in federally funded programs in 2005-2006?

    c. Describe the trend that you notice in the histogram.

    d. How is the histogram you constructed different from a bar graph?

    e. What is the percent of increase in special education enrollment from 1994–95 to 2004–05?

    f. To what might the increase in enrollment be attributed?

    g. What happened to enrollment from 2004-05 to 2005-06? What do you think caused this change in enrollment?

    h. Why do you think providing educational opportunities for students with special needs is an example of social justice?

    i. What questions do you have about the data?

3. Write a letter to the editor of any newspaper describing the trends in the data and the continued need for attention to social justice.

Appendix 25

Here's the Graph—What's the Story?

Name _____ Date _____

School _____ Class _____

DIRECTIONS: For each graph, think of a situation that the graph could represent.

1. Determine a possible topic and title for the graph and label the axes appropriately.
2. Determine whether the graph needs a key.
3. Write a story about each graph, describing the data that could have been used to create the graph.

Title: _____

Title: _____

Graph 1

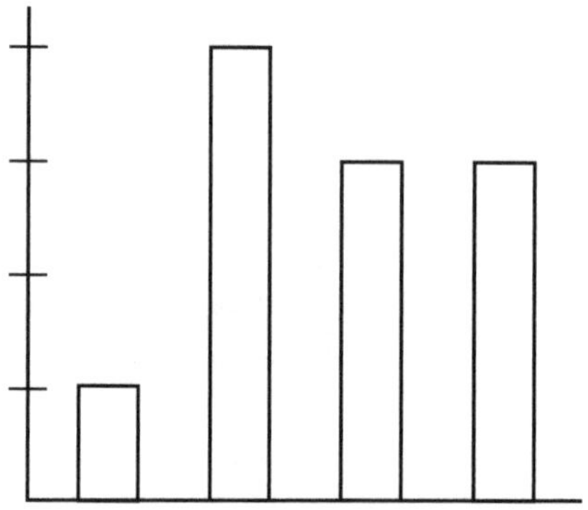

Graph 2

Title: _____

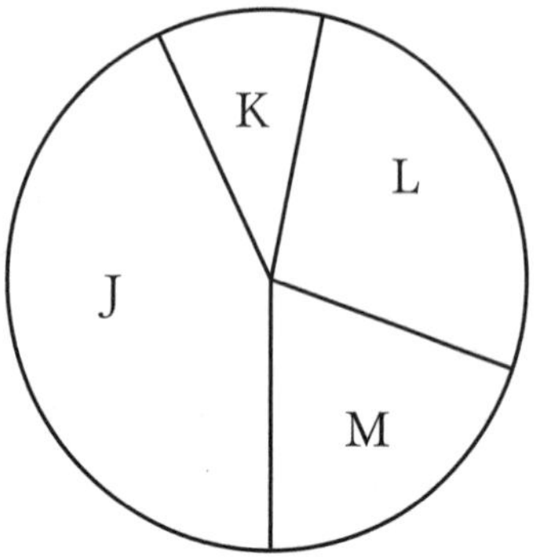

Title: _____

Graph 3

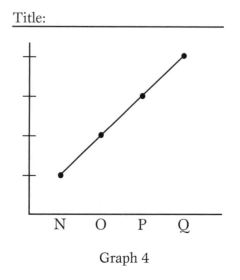

Graph 4

Here's the Graph—What's the Story?

Name _____   Date _____
School _____   Class _____

DIRECTIONS: For each graph, think of a situation that the graph could represent.

1. Determine a possible topic and title for the graph and label the axes appropriately.
2. Determine whether the graph needs a key.
3. Write a story about each graph, describing the data that could have been used to create the graph.

Title: _____   Title: _____

Graph 5

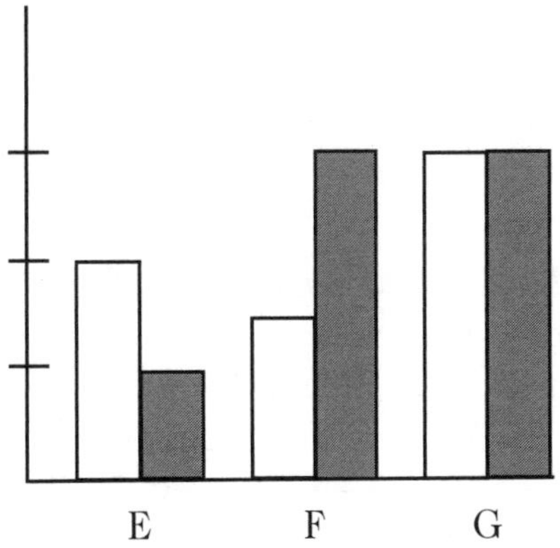

Graph 6

Title: _____

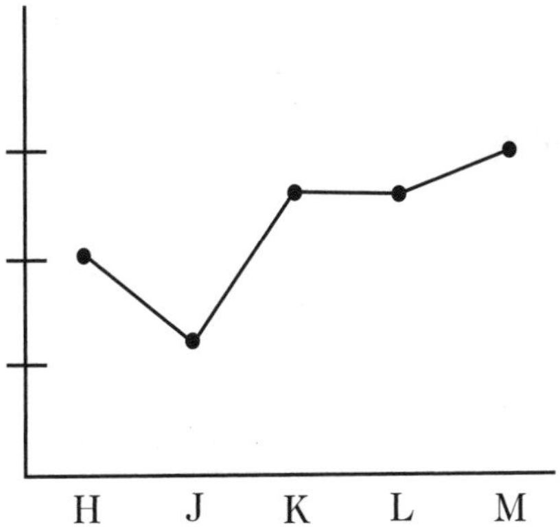

Graph 7

Appendix 26

Here's the Story—Where's the Graph?

Name _____   Date _____

School _____   Class _____

DIRECTIONS: For each story or set of data, think of what a graph representing the story or the data might look like.

1. Sketch a graph and label all the essential elements.

2. Write an explanation about what clues in the story or in the data made you sketch the graph the way you did.

Story #1: Janelle is in the sixth grade. To earn her weekly allowance, Janelle does a lot of errands around the house. She spends 1/3 of her weekly allowance on lunch, 1/4 of her allowance on carfare, 1/6 of her allowance on CDs, and the rest she saves in her bank. Sketch a graph to represent this story. (How much money might Janelle receive for her weekly allowance?) What clues in the story helped you to sketch the graph?

Story #2: After taking a survey of the number of televisions in each home of the seventy-two fifth graders in the Academic Elementary School, the students found that about 75 percent of the students had a television in the living room, about 20 percent of the students had a television in the family room, about 16 percent of the students had a television in the kitchen, and about 15 percent of the students had a television in their bedroom. Sketch a graph to represent this story. What clues in the story helped you to sketch the graph?

Data Set #1: A physician told Tashon's mother to watch her diet. She weighed 64 kilograms in the physician's office. During the ten weeks that she adjusted her diet to reduce her intake of fat and include more vegetables and fruit, Tashon decided to help her mother record her weight. Here's what Tashon recorded:

After	2 weeks	4 weeks	6 weeks	8 weeks	10 weeks
Weight in kg	60 kg	58 kg	58 kg	56 kg	55 kg

Sketch a graph to represent the data. What clues in the data helped you to sketch the graph?

Data Set #2: After returning from shopping for groceries with his father, Sam examined the receipt: about $18 for vegetables, about $9 for fruit, about $3 for paper goods, and about $6 for cat food. Sketch a graph to represent these data. What clues in the data helped you to sketch the graph?

Data Set #3: During different periods of time, different buildings were constructed to accommodate large numbers of people. The heights of four buildings are listed below, along with the dates they were built. Sketch a graph to represent these data. What clues in the story helped you to sketch the graph?

Building	Height	Date Completed
Burj Khalifa, Dubai	2,717 feet	2010
Patronas Tower, Kuala Lumpur	1,483 feet	1997
Empire State Building, USA	1,250 feet	1931
St. Peter's Basilica, Italy	461 feet	1590

(from Ash 1996, p. 48; Thomas, 2010, p. B1)

Appendix 27

Graphs and Titles—Where's the Match?

Name _____   Date _____

School _____   Class _____

DIRECTIONS: Find the title that you think is appropriate for each graph.

1. Match graphs and titles.
2. Write an explanation about what clues in each graph helped you to match the graph with the title.

GRAPHS

TITLES

1.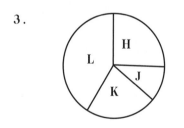

(a) Types of Footwear Worn by Students in the Sixth Grade

(b) The Relationship between Height and Weight

(c) Number of Sit-ups Students Can Do in One Minute

(d) The Strength of a Battery over Time

(e) Favorite Ice-Cream Flavors of Sixth Grade Students in Our School

(f) Number of Times a Vowel Appears in One Newspaper Article

(g) Number of Riders in Cars Passing Our School between 10:00 and 11:00 a.m.

(h) Types of Food We Eat for Lunch

2.

3.

4.

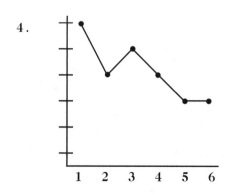

Appendix 28

Name _____   Date _____

School _____   Class _____

"Not Eating Your Veggies? It's No Joke"

Your Health
By Kim Painter

For kids to be persuaded, adults need to get serious

In 2006, comedian Jay Leno proudly told *Parade* magazine that he hadn't eaten a vegetable since 1969. And it looks as if he is sticking to his no-green-food ways: In a recent skit, he accompanied Pee-wee Herman (aka Paul Reubens) to a salad bar—and declined everything except a cookie and a batch of deep-fried potatoes (technically a vegetable but not exactly a health food).

While Leno may be a vegetable-hating extremist, he's hardly alone in failing to get enough produce in his diet. Just 27% of adults in the USA eat at least three servings of vegetables a day, and just 33% eat at least two servings of fruit a day, said a report released in September by the federal Centers for Disease Control and Prevention. A scant 14% meet those marks for both fruits and vegetables.

So it is time to give up on selling broccoli to reluctant grown-ups like Leno and just focus on kids?

No, insist nutrition and public health experts. "It is absolutely possible for adults to change their eating patterns," says Marisa Moore, a registered dietitian in Atlanta and spokeswoman for the American Dietetic Association. "And adults are the best role models when it comes to children." They also buy most of the food. So for the good of both adults and children, she says, "we can't give up on them."

Adults do change, says Heidi Blanck, senior nutrition scientist at the CDC: "Older adults do have higher (produce) consumption rates than younger adults. We see it start to creep up right around age 45."

That's about the time many people develop health problems linked to poor diets, she notes. But ideally, people would discover the joys of a crisp apple and a delicious veggie stir-fry long before their clogged arteries screamed for mercy.

For years, the federal government tried to reach consumers with its "5 Day" campaign. In 2007, it was replaced with "Fruits & Veggies: More Matters," a partnership led by CDC and the private Produce for Better Health Foundation.

The name change reflects new thinking: Experts say individual needs vary, but in general, more is better. (You can get personal recommendations at fruitsandveggiesmatter.gov.)

A good rule of thumb, says foundation president Elizabeth Pivonka: Half of what you eat should be fruits, vegetables or beans.

But if you haven't heard of "More Matters," Pivonka isn't surprised. The program runs on just $4 million a year. That's a tiny sum in the world of food marketing, where big bucks are spent to sell burgers, chips and soda. For now, she says, the campaign is a "slow, steady effort" that depends on supermarkets and others to spread the word.

Meanwhile, some are trying other tactics, including taking fruits and vegetables to workplaces. At the Texas Department of State Health Services, employees can order weekly baskets of local produce to be delivered to their offices, Blanck says. People who work for the CDC in Atlanta can shop at weekly parking-lot farmers markets. Health care giant Kaiser Permanente has a similar program.

At 12 Kaiser locations with weekly farmers markets, 77% of repeat shoppers say they are eating more fruits and vegetables, says Preston Maring, a Kaiser physician who leads the effort. People can't "pass up a fresh peach in the middle of July or a fresh spear of asparagus in February," he says.

Well, maybe a certain comedian could. But for others, there's hope.

Source: USA TODAY, 19 October 2009.
Reprinted with Permission